The Genius of Glenlair:
James Clerk Maxwell

Portrait by his cousin, Jemima Wedderburn

a reader-friendly biography
by Ellen Johnston McHenry

ISBN: 979-8-9868637-9-5

For bulk orders by retailers, please inquire at IngramContent.com

The plaid pattern on the cover is the Maxwell tartan.

NOTE ABOUT SUGGESTED ACTIVITIES:

You will see superscripts like these: $^{(*1.2)}$ at various places in the text. They indicate that there is a suggested activity that corresponds to what you just read. The activities are listed in appendix 4. You don't have to do the activity at that point; you can wait and do all the activities when you reach the end of the chapter, or the end of the book. Or you can just read the biography and skip the activities altogether. However, the superscripted notes are there in case you decide to punctuate your reading with activities.

TABLE OF CONTENTS

CHAPTER ONE

James was born on June 13, 1831, in Edinburgh *(ED-in-bur-uh)*, Scotland. His parents, John and Frances, came from families that had inherited quite a bit of farm land. They were not rich, but they were far from poor. The family name had originally been just "Clerk" (which in Scotland is pronounced "Clark"). John had been obliged to add "Maxwell" to his surname in order inherit some property. Someone many generations back had put a legal requirement on that land, stating that whoever owned that land must be named Maxwell. If your name wasn't Maxwell, you'd have to change it. So even though John was the rightful heir, he still had to change his name from "Clerk" to "Clerk Maxwell."

John Clerk Maxwell's property included a manor house called Glenlair. When James was about 3 years old, John decided to move the family from their house in the city at 14 India Street (which is now a museum) out to Glenlair. There were many acres of farm land around the house, as well as areas of woods and streams. It was at Glenlair that James spent most of his childhood.

Glenlair in 1884

1

Who lives in Glenlair now?

Glenlair was sold several times after James's death. The current owner is Duncan Ferguson, whose father bought the property in 1950. The manor house had almost burned to the ground in 1929, so it need a lot of renovation. Many things were updated to make the house comfortable for modern living (indoor plumbing, for example). Duncan Ferguson realized that his property had significant historical value, and has begun to restore as much of the property as possible to the way was during James's childhood. Many trees have been planted, and farm buildings have been "backdated" to look like they did during James's childhood. These backdated buildings are for educational purposes, to show visitors what farming life was like in the early 1800s.

A painting from the 1830s of James with his mother, Frances

All children are curious, but James was exceptionally curious. He wanted to know exactly how everything worked. His mother wrote that he was always asking, "What is the go of that?" (Meaning, "What makes it go?") If the answer did not satisfy his curiosity, he would then ask, "But what is the particular go of it?" He would pester not only his parents, but every hired hand on the property, wanting to know all the details about every aspect of farm life. He helped with farm chores outside, and with indoor activities like basket making, knitting, and baking.

One day, little James was given a metal pie pan to play with. As he turned it over in his hands, he noticed that sunlight was being reflected off the pan and onto the ceiling. This delighted him to no end and he ran to tell his mother that he had found a way to bring the sun inside. [*1.1]

James grew up as an only child, but he had several cousins that he was close to. His cousin, Jemima, was a bit older, but close enough in age that they spent many happy hours playing together. They were both artistic, and Jemima grew up to be a professional artist. The drawings shown here on the right were done by Jemima herself, documenting a little bit of James's childhood. On the top you see them paddling in tiny boats. We know that they went boating in Glenlair's pond, although the location in the watercolor painting could be somewhere else, like a local creek.

A zoetrope from the 1800s

James and his cousins also enjoyed one of the latest toys, a zoetrope *(ZOE-eh-trope)*. James and Jemima would draw their own picture strips to put into their zoetrope. The zoetrope was one of the earliest forms of animation. It is one of the forerunners of movies. Picture strips of birds flying, horses running, or people dancing were ideal for zoetropes because these motions are repetitive. The circular picture strip went around and around, showing the pictures over and over again. [*1.2]

James learned to read very quickly. His parents chose to keep him at home instead of sending him to school, so that he would have plenty of time to read on his own. They had a good collection of books in the house and James read a lot of history, geography and literature, including the Bible, Shakespeare and John Milton. He was also good at memorizing, and by the age of 8 he had memorized the entire 119th Psalm (which is 176 verses long) and other passages of Scripture and the words to many hymns. Religion was important to the family, and

James loved dogs. The family dog was always named Toby.

they not only went to church every Sunday, but gathered the whole household for prayer every day. (This was not unusual for families in the 1800s.)

In an age without radio or television, family entertainment would often include reading novels or poetry out loud, or acting out scenes from a play. James developed an appreciation for poetry and would often write poems in later life. The Maxwells entertained guests frequently, so James learned how to behave properly at fancy dinners. There were also many community events to go to, such as fairs and dances. James gained many important social skills that would come in handy later in life.

James's childhood was idyllic until the tragic death of his mother from stomach cancer when he was 8 years old. (Sadly, years later, James would die at the same age as his mother, and of the very same disease.) Even after her death, life at Glenlair continued to be good, but James missed his mother deeply. James's parents had intended to teach him at home until he was 13, then send him to the university. In those days, teenagers sometimes went off to college if there was not a suitable secondary school nearby. John dreaded the loneliness he would experience if James was sent away to school, but as a busy lawyer he did not have the time to teach James himself. So John decided to hire a tutor.

The only suitable person in the neighborhood was a 16-year-old boy who had just finished his education and done very well on all his final exams. It was not uncommon back then to hire teenagers as tutors for younger children, so this decision seemed to make sense. This young tutor turned out to be a disaster, however. The tutor was harsh with James and made him memorize and recite many things that James thought were pointless. If James made a mistake, the tutor would give him a hard smack.

Doubtless, the tutor had himself been treated this way in school and thought that this was the correct way to teach. James hated this method of schooling and after putting up with it for over a year, he could finally take it no longer. One day, right in the middle of a lesson, 10-year-old James ran out of the house and got into one of the tubs that he and Jemima used as boats, and paddled out into the middle of the duck pond. The tutor shouted and threatened from the shore, but James would not come back. This incident made James's father realize how much James disliked the situation. The tutor was sent away and James went back to learning on his own. The drawing below was done by cousin Jemima. Notice the date in the corner: 1841. The pond was much larger than shown here but if she had made the pond and the people the right size, the people would have appeared too small on the page.

When James was about 13, his Aunt Jane (his mother's younger sister) thought it was high time that James was sent to school, so she proposed a plan. Another aunt, Isabella, (Jermima's mother) lived just down the street from the Edinburgh Academy. James could stay with her during the school terms then return home during the summer and during holidays. John agreed that James was now old enough to need excellent teachers, so he agreed to the plan.

The Edinburgh Academy was like a private high school. Not everyone went to high school in those days.

When James arrived for his first day of school, he was dressed in his farm clothes, which were warm and comfortable, but not very fashionable. The older boys immediately saw James as a target for teasing. After a rough start at school, James's aunts stepped in and bought him a wardrobe of proper city clothes so he would not look different. They could not do anything about his country accent, however, which had earned him the nickname "Daftie" from his classmates. During recess, James would often wander off by himself to observe insects or climb trees.

Even in this improved school environment, James still found classroom learning dull compared to reading books from his aunt's library. School education consisted mainly of memorizing Greek and Latin words, and doing arithmetic, neither of which James liked. However, he managed to win a prize at the end of the year, getting the top award in the category of Scripture biography (people of the Bible). He got this prize again in his second year, also. As the third year began, he was introduced to geometry, a subject he immediately liked and was very good at.

One Saturday, James's father came to visit him in Edinburgh. He took James to an exhibition of new inventions called "electromagnetic devices." The early 1830s had seen Michael Faraday discover that there is a connection between electric current and magnetism. He had produced magnetism from electricity, and electricity from magnetism. In 1833, when James was just 2 years old, a practical use was found for temporary "electromagnets" that could be turned on and off: they were used to make a communication device called a telegraph. Samuel Morse wrote a dot and dash code that could be used to transmit words over an electric wire. Soon telegraph wires stretched from city to city. James was very interested in how these machines worked. Later in his life he would help Faraday complete the theory of electromagnetism.

An early telegraph machine

When James was 13, he discovered geometric polyhedra. He made a tetrahedron (4-sided figure) and a dodecahedron (12-sided figure) out of cardboard and mentioned them in a letter that he wrote to his father. He learned that there are only five regular polyhedra (with number of sides 6, 8, 10, 12 and 20) but these can be used to make other more complex shapes. From this time on, mathematics began to interest him more and more. [*1.3]

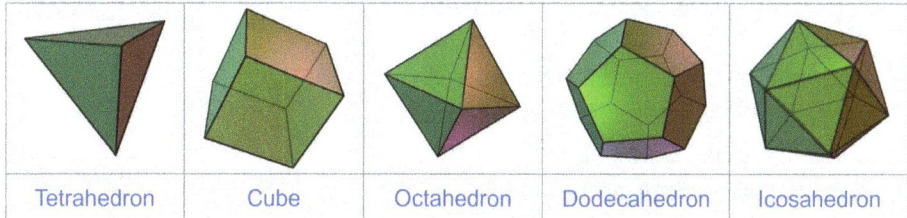

| Tetrahedron | Cube | Octahedron | Dodecahedron | Icosahedron |

Also at this time, James made his first lifetime friend. One of his classmates, Lewis Campbell, moved with his family to a house almost next door to where James lived with Aunt Isabella. James could not only walk to and from school with Lewis, but spend after school time with him as well. Lewis was one of the top students in the class, and James was thrilled to have someone his own age who shared his interests and abilities. As an adult, Lewis would become quite a scholar, writing books about ancient Greek philosophers. Later in life, Lewis would write a biography of his friend, James.

Lewis introduced him to another friend, Peter Tait, who would not only become another lifetime friend, but would also become one of Britain's top physicists of that century. James and Peter would challenge each other with mathematical puzzles. One puzzle they never found an answer to was how to make a mirror that would not reverse your face. (Do you think this is possible?)

When James was 14, he became almost obsessed with drawing ellipses. An ellipse is sort of a flattened circle. James found that he could draw an almost perfect ellipse by using two pins, a piece of string, and a pencil. He tied a piece of string loosely between the pins, then pressed the pencil against the string and moved it along, keeping the string tight. If he made the pins close together, he got a very round ellipse. If he put the pins far apart, he got a long, thin ellipse.

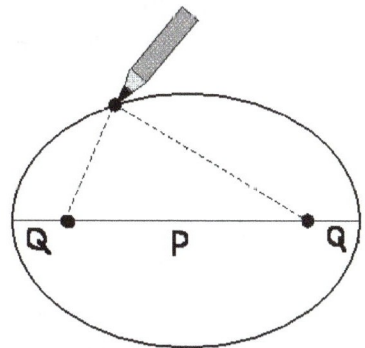

He discovered ways to use math to describe what he was doing. "S=P+2Q" says that the length of the string, S, is equal to the distance between the two pins, P, plus twice the distance Q, between a pin and the edge.

Then he started thinking about ways he could continue to experiment. He untied one end of the string from the pin and put it on the pencil. He looped the string round the other pin, then pushed it tight against the top part of the string. This arrangement made an ellipse-like shape called a Cartesian oval. A mathematician named Rene Descartes had discovered this shape back in the 1600s. James was probably unaware of Descartes' work at this point. Quite on his own, James decided to figure out a way to explain an ellipse with a math equation.

After succeeding in that, he then experimented with adding a third pin. He went on to experiment with four pins, then five. He took notes on each experiment and began to find ways to summarize all of his findings with math formulas.

When James showed his father what he had been working on, his father was very impressed. He took James's papers to a friend who was a professor at Edinburgh University. The professor was also very impressed and realized that James's work could be turned into a published paper. Several math professors then helped James to polish his theories and get them ready to submit to a printer. One of the professors read the paper to the Royal Society of Edinburgh. (The Royal Society was an exclusive "club" where the very best scientists and mathematicians of the day would share their ideas.) [*1.4]

After the publication of the ellipse paper, James became interested in the writings of Descartes. After Descartes, he went on to read other mathematicians and scientists of previous centuries. He felt a kinship with them, almost as if they were his friends. He didn't idolize them, however, and was very aware that even

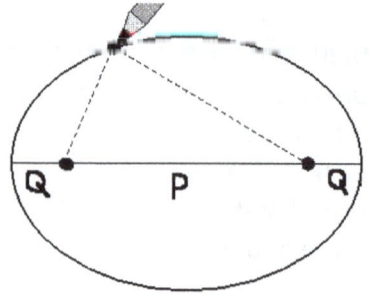

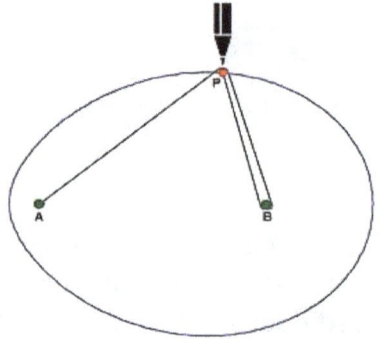

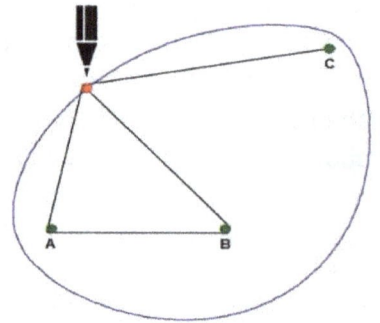

these great men had made mistakes. (He had found a small error in one of Descartes' calculations.) James knew that he, himself, sometimes made mistakes in his equations and calculations, and resolved never to become prideful about his work. He would keep this promise to himself and people would later remark how amazingly humble this great man was.

He was in many ways still a playful boy in his first year or two at the academy. His letters to this father during this time often included silly jokes. He thought it was fun to mix up the letters in his name and make an anagram. For instance, once he signed his letter as "Jas Alex McMerk-well." Once he even included the mailman in his jokes and wrote the address of Glenlair as "Postyknowswhere." (*1.5)

His last years at the Edinburgh Academy were very successful. Not only did he do well in math, he also won prizes in the categories of history, geography and French. He wrote poems for fun and continued to read great literature. When school was not in session, James was at home at Glenlair, helping his father to maintain the house, the barns and the land. No job was too small for him to attend to, and no employee of the farm was unimportant. James was a helper and friend to all. His interest in farming, and his willingness to participate in the labor that farming required, continued throughout his life. This made him unusual among his peers. Rarely did highly educated men help with manual labor of any kind.

Meanwhile, James's cousin, Jemima, had grown up, and was now married to a mathematician named Hugh Blackburn, a professor at Glasgow University. This painting (by Jemima) shows Hugh taking her picture with a camera. Photography had been recently invented;

James would some day make a contribution to the development of color photography.

James would often go to visit Hugh and Jemima, and on one of these visits James was introduced to the youngest professor at Glasgow, William Thomson, who was only 22 at the time. James and William struck up a friendship that would last for the rest of their lives. Thomson would eventually become "Lord Kelvin," the person for whom the Kelvin temperature scale was named. (Kelvin degrees are similar to Celsius degrees.)

After graduation from the academy at age 16, James was ready to begin studying at Edinburgh University. His father wanted him to keep up the family tradition and study law. This might sound to us like his father was being selfish or unreasonable, but he really wasn't. Back then, you couldn't study science in the way that you can today. The word "science" wasn't even in use yet; scientists were called "natural philosophers." Natural philosophy was seen as more of a hobby than an employment, and the people who dabbled in the natural sciences were either clergy in the church who didn't mind being poor, or they were people who were born wealthy and had lots of time and money to spend on their hobbies. The Maxwell family wasn't poor, but they did need to have jobs that brought in a decent income. James would have to find a career that paid well. James had proved himself to be very well-rounded in his abilities, excelling in all subjects, not just math. He would be able to do anything he set his mind to. Thus, James went off to the university, thinking that we would be a lawyer some day, just like his father.

Edinburgh University *(remember, it's pronounced "ED-in-bur-uh")*

CHAPTER TWO

At Edinburgh University, it was back to studying history, geography, Greek, Latin, and philosophy. These subjects were considered to be foundational for any career, whether it be in law, medicine, theology or teaching. His classes explored many philosophical questions and James enjoyed the lively discussion and debates among the students. Philosophers at this time were starting to ask questions like, "How can we know that we know something?" and "Can the tools of logic be applied to philosophy and theology (the study of God)?" A German philosopher named Immanuel Kant had just proposed that all knowledge is relative, and the only thing we can know about something is its relationship to other things. James thought about this deeply and, without coming to any conclusions about whether or not it was true in a philosophical sense, explored how this might be applied to material things—to matter itself (the "stuff" that everything is made of). James wrote in an essay for his class, "The only thing that can be directly perceived by the senses is Force, to which may be reduced light, heat, electricity, sound and all the other things which can be perceived by the senses." We see light that bounces off objects, we hear sound waves that some things produce, we see the needle on the scale go up when we weigh something, we feel sensations in the nerves of our skin when we hold things, but can we actually come into contact with the essential stuff that things are made of?

Today, we know that matter is made of atoms, and that atoms are made of a nucleus surrounded by layers of fast-moving electrons. The outermost "shell" of electrons will repel other shells, keeping atoms from blending together or overlapping. Atoms can and do join together to make molecules, sometimes by borrowing or sharing a few outer electrons, but the electrons keep the central part—the nucleus—away from the nuclei of other atoms. The nucleus of an atom (and the inner rings of electrons) never touch any other atom. Your foot doesn't

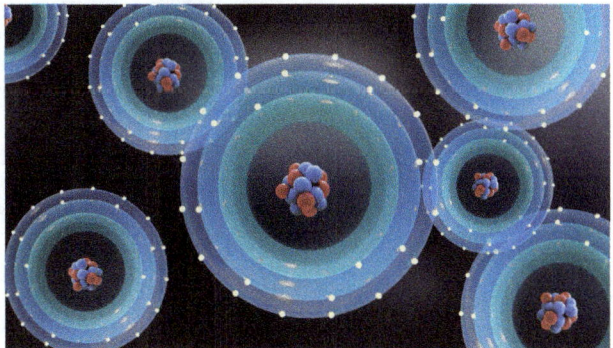

go through the floor because the outer layers of electrons in the atoms that make the floor are pushing against the outer layer of electrons in the atoms of your foot. So do things actually "touch" each other? It depends on your definition of "touching." Since electrons had not yet been discovered when James wrote his essay, we don't know exactly what was going through his mind. Most importantly, his essay gave a hint of what a deep thinker he would turn out to be.

There was another side to James's academic life at Edinburgh University. Classes were mostly about thinking and writing, but one professor, James Forbes, taught his students by using experiments and what are now known

Professor James Forbes

as "hands-on" demonstrations. He encouraged his students to get their hands dirty and try things to see if they worked. He allowed James to have complete access to the laboratory and try any experiment he wanted to do. James learned to use all the tools and machines in the lab. This method of teaching would someday become his own method of teaching. As a professor, James would say, "Never discourage someone from trying an experiment; if they don't find what they are looking for, they may find something else just as important."

James Forbes was very much like James Maxwell, and had a great sense of adventure. He would take students on hiking trips to explore natural landmarks. Forbes was interested in geology and was a pioneer in the fields of seismology (sound waves in the crust of the earth) and the study of glaciers. Forbes was also a good writer and would help James to edit his papers to make them more understandable to the readers. Forbes was a great source of inspiration to James and shaped his life significantly.

Hikers in Scotland

James would do a lot of teaching in future years. He believed in exposing children to science at a young age. He would recommend that if any child has an interest in science, a visit to a "real man of science in his laboratory" can be a turning point. Don't worry that the child can't understand all the explanations, he would say. It is more important to see how a "man of science" does not get angry when something goes wrong, but patiently tries to figure out why he got those results. (This illustration shows what labs looked like in James's day.)

In the 1800s (and even in the early 1900s), many families lived on small farms and were dependent on the crops and animal products they produced during the year. Edinburgh University took great pride in the fact that they had many young folk from farming families enrolled as students. They believed in education for all, no matter how rich or poor a family was. The university would close from the end of April until the end of October so that students could go home and help on their family farms. So James found himself back at Glenlair only half a year after he started. He was expected to keep up his learning at home during the summer, and was given a supply of books from the school library. He also brought home some lab supplies so that he could continue experimenting. He set up a temporary workshop above the wash house, and used an old door and some barrels as his workbench. There was a panel in the roof that would slide open to

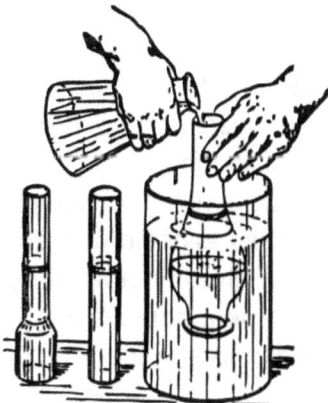

let light in and fumes out. In a letter to Lewis Campbell he described his supplies: "bowls, jugs, plates, jam jars, water, salt, soda, sulphuric acid, broken glass, copper and zinc plates, bees' wax, clay, rosin, charcoal, a lens, a Galvanic apparatus *(the battery of that day)*, and a countless variety of little beetles, spiders, and woodlice that fall into different liquids and poison themselves." James's family called his lab "Jamesie's dirt."

As much as he liked chemistry and electricity, the experiments that interested James the most at this point in his life were the ones that used polarized light. Of course, no one knew yet what light really was. It would be James himself who would figure that out in later years. But much was known about how light behaves. Isaac Newton had shown that light is made of a complete spectrum of colors. Light can be bent by a prism, with each color bending a different amount, allowing the colors to be displayed in a band that we call a rainbow. (By using another prism, he put the colors back together again to make white light, clearly showing that the colors were in the light, not in the prism.) Lenses had been used for centuries to bend light in ways that would enlarge or shrink images. Telescopes and microscopes had been in use for well over a century by James's time. More recently, new discoveries had been made about polarized light. [*2.1]

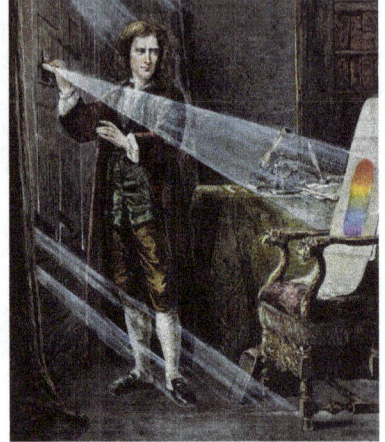

Isaac Newton used prisms to discover that sunlight was made of many colors.

James knew that when light reflected off a surface, sometimes the reflected rays of light would cause strange effects. We call this strange effect "polarized" light. When polarized light is used to look at transparent (clear) objects, sometimes beautiful swirly color patterns are visible. Exactly why this happened was unknown, and James wanted to learn whatever he could about polarized light. He remembered the large pieces of Icelandic spar (a clear mineral) that he had seen in the collection of one of his professors but knew there was no way he could afford to buy a specimen like those. Instead, he tried gluing pieces of glass inside a matchbox. He also got pieces of a much less expensive mineral (mica) and spent hours slicing and polishing them.

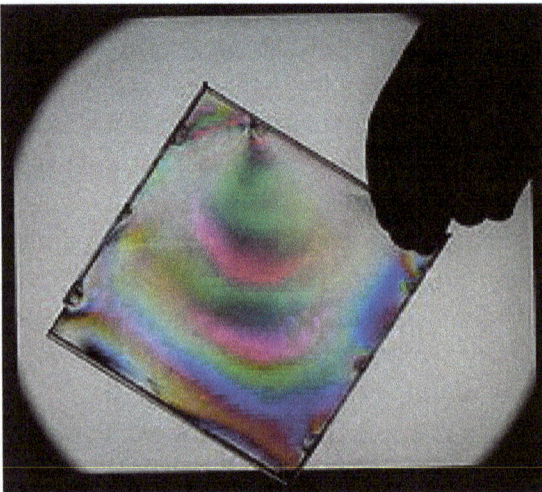

Polarized light can show patterns in clear plastic objects.

He had quite a bit of success even with these less-than-perfect polarizers. He looked at melted glass through his polarizers and realized that the colored rings he was seeing weren't just pretty patterns, they were actually showing him something important. Those colored lines showed him where the stressed areas in the material were. To test this theory, he knew that he should use a material that could be stressed while he was watching it under the polarizers, so that he could see where rings and lines appeared as he applied stress. He needed something that was clear but also soft enough to bend and stretch. He decided to head to the kitchen to cook up a batch of clear jelly. He made a ring of jelly with a cork stuck in the center. He would twist the cork to put stress on the jelly but not actually tear it. Imagine someone putting a cork in your belly button then twisting it a bit. Ouch! That was what James was doing to the jelly. There was a colored ring or line in the places that were going "ouch."

He even found a way to record the things that we was seeing. He made a simple device called a camera lucida, that projected the image onto a piece of paper. He would then use watercolor paints to trace the image onto the paper. He sent some of his watercolors to an artist he knew who was also very interested in polarized light. The artist was so impressed with James's work that he sent him two large specimens of Icelandic spar. These mineral specimens became some of James's most treasured possessions. (*2.2)

A very nice specimen of Icelandic spar

The long holidays from school also gave James opportunities to spend time thinking about geometry problems. He became interested in cycloids, the line patterns you get when you roll a circle across a surface. For example, if you put a dot on a bicycle wheel, then follow the dot as the bicycle rolls along, you get a line that looks like a series of half circles. He wrote a paper about the equations he had come up with to describe

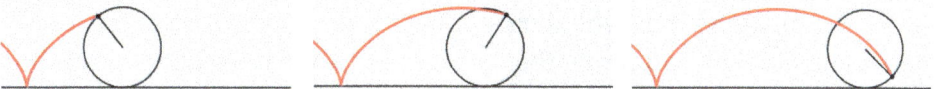

these shapes, and it was presented at a meeting of the Royal Society. He was getting very good at using math to describe things. He even found a way to use math to describe the colorful patterns that he saw in the jellies when he put them under the polarized light. He wrote up another paper for the Royal Society. This time, however, he was sloppy in his writing and his professor, James Forbes, told him to rewrite it because he was capable of much better. James never forgot this hard lesson and from that time on he took the time to carefully edit his papers. (*2.3)

Eventually, James's father realized that James wasn't going to be a lawyer. Any young man who can have three mathematics papers presented to the Royal Society before he turned 20 deserved to have an opportunity to study math and science at a world-famous university. Therefore, it was arranged for James to attend Cambridge University, north of London, England. Cambridge had produced many fine mathematicians and scientists, including Isaac Newton, Francis Bacon, William Harvey, Charles Babbage and Henry Cavendish. The name James Clerk Maxwell would soon be added to that list.

The schedule at Cambridge was more intense than anything he had experienced before. His classes really kept him busy and he found it difficult to accomplish everything he wanted to do. Sometimes he had to do things at odd times of the day. For example, he tried jogging after he finished his studies at about 2:00 AM. His dormitory had two floors with a staircase at either end and he would use them as a running track. The students who lived in the rooms off these hallways were less than pleased to have someone running past their room every few minutes in the middle of the night. So they would hide behind their doors with objects like boots and brushes and throw these at him as he went past. James got the hint and decided maybe he should change his schedule a bit.

Soon after he arrived at Cambridge (in the section called Trinity College) he accepted an invitation to join an exclusive club called the Selective Essay Club. There were only 12 people at a time in this club, and because of this they sometimes called themselves The Apostles Club. They would take turns reading essays they had written and then discuss and debate them. James loved to think and write about philosophical issues, not just mathematics, so this gave him an opportunity to do so. One of his essays was called "Is Autobiography Possible?" An autobiography is when someone writes their own life story. James stated his belief that no one should reveal to the public the private details of their life. (What would he think of Facebook and Twitter?!) We know many details of James's life because they were written down by his friend, Lewis Campbell, or by his cousin, Jemima.

James missed all the animals back at Glenlair, especially the dogs. (We know that at least one of the dogs was a white terrier.) There were some cats in the building where he lived at Cambridge, probably kept there to keep the rodent population down, so James befriended them and allowed them to come in his room. Some sources say that he did experiments on them, dropping them from various heights to see if they landed on their feet. Later in his life, James denied ever having "tortured" cats. He did admit to dropping them from a safe distance, only a few feet off the ground. He was more

Falling cats always land on their feet!

interested in how they would respond to a very short drop than a long one.[*2.4]

Because James spent so much time studying while at Cambridge, he didn't have time to do many of his own investigations. Besides all the regular exams, he also entered some very tough competitions. The most difficult one lasted seven days! James was really good at math, but there was actually a student in his class who was a little bit better than he was, so James came in second in the math exam. Second was very good, though, and he came in first in another category. He did so well, in fact, that Cambridge gave him an offer to stay on as part of their staff, teaching the beginning students. After a few years, he would then be qualified to apply for a job as a professor at any university he chose.

Even though James was very busy with his own studies, he always found time to help others. We know that he once helped a friend who had eye trouble by reading the lessons to him. He visited friends who were sick and tried to cheer up those who were discouraged. He helped first-year students who were struggling and needed extra tutoring. Someone who knew James during his days at Cambridge said this about him: "He was friendly and kind. Everyone who knew him can recall some kindness or some act of his which left a permanent impression of his goodness on their memory, for Maxwell was "good" in the best sense of that word."

After graduation, James had more time to spend on his own experiments. He did have teaching duties, but evenings and weekends allowed plenty of time to continue his scientific interests. In addition to his teaching and experimenting, he also spent more time exercising.

The River Cam next to Cambridge

He would go on hikes, swim or row boats in the Cam river, or work out in the new gym on campus. He even helped to organize a swimming club.

James found himself thinking a lot about light and color. He'd been thinking about color ever since he was a toddler, when he would ask people why the sky was blue and no one could give him a satisfying explanation. Other things puzzled him as an adult. Everyone knew that sunlight contained all the colors of the rainbow, but what about the colors brown and gray? They weren't in the rainbow. Did light combine in the same way that paints combine? Artists mix blue and yellow paint to make green paint. Did that mean that yellow light and blue light make green light? And what role did the eye play in seeing light? Very little was known about the inside of the eye.

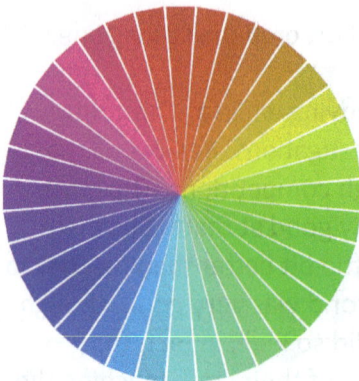

18

James decided to learn everything he could about eyes. He made a device that would allow him to look inside an eye. He had a very hard time convincing anyone to let him stare at their eyes, so he started by using a mirror to look in his own eye. Then he realized that animal eyes were very much like human eyes, and he knew several dogs in the area. The dogs might have been shy or squirmy at first, but James was an excellent dog trainer and after a few lessons and some yummy treats, the dogs would allow him to stare into their eyes and make observations. James wrote about what he saw in dog eyes: "They are very beautiful, with glorious bright patches and networks of blue, yellow and green, with blood vessels great and small."

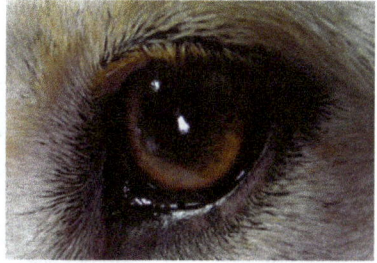

Finally, he found a few people who trusted him enough to let him use his tool to look at the back of their eyes. He thought that human eyes were not nearly as beautiful as dog eyes, but humans could use words to tell you what they were seeing, and that was helpful sometimes. He saw that the back of the eye, the retina, was fairly dark, with a small yellowish spot near the center. In

The back of the eye is called the retina.

this picture, the yellow spot is the optic disc, the place where nerves and blood vessels come into the eye. What James did not know is that this optic disc creates a "blind spot" in our vision. We don't notice this blind spot because our brain fills in the missing details. [* 2.5]

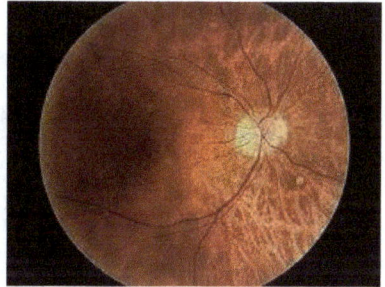

At the same time, James decided to begin experimenting with color. He remembered an experiment that Professor Forbes had done, putting colors onto a spinning top. Prof. Forbes had explained that the colors spin so quickly that the eye can't react to each individual color but combines them into one color. So James made his own version of Forbes' spinning color top. The wheel shown here is a photograph of the actual wheel that Maxwell made.

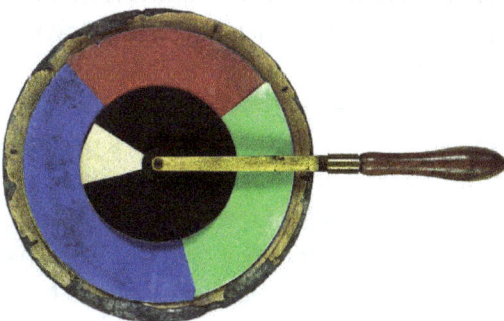

Thomas Young

James Maxwell holding his color wheel

Before we go on, we need to give credit where credit Is due. Professor Forbes's ideas about light and color came directly from someone else's work. Professor Forbes had read about the theories and experiments of Thomas Young. Young was still alive while Professor Forbes was growing up, but had died two years before James was born. Interestingly, James and Thomas shared their birth date, June 13. They also shared the trait of being interested in a lot of topics, and being good at just about anything they set their mind to. By the age of 14, Thomas knew Latin and Greek and was familiar with many other languages such as Hebrew, Arabic, Turkish and Persian. Later in life he would study Egyptian hieroglyphic writing and help to decipher the famous Rosetta Stone.

Like James, Thomas studied at both Edinburgh and Cambridge, although he went other places, too. Unlike James, his university degree was in medicine, so his daily job was being a doctor. In his spare time he studied other things, such as languages, math, music and the science of light. On the basis of his experiments comparing water ripples to light ripples, Thomas had concluded that light must be a wave, not individual particles. This contradicted the writings of many scientists including Isaac Newton. Newton believed that light was made of invisible particles that moved very fast. Most scientists of Thomas Young's day agreed with Newton. In the end both turned out to be correct, as light can act like both a wave and a particle.

Also like James, Thomas was interested not only in light itself, but also in the eye as the receiver of light. Thomas believed that the eye had only three basic color receptors; all other colors were the result of how the eye and the brain perceived mixtures of these three. He turned out to be right, but how he came up with the number three is unknown.

James decided to pick up where Thomas had left off and try to figure out the nature of those three receptors. He made three paper disks: red, blue and yellow. He knew these were the "primary" colors that artists used. He knew that blue and yellow paint can be mixed to make green, so he made his wheel half blue and half yellow and gave it a spin. What he saw on the spinning disk was not green, but pink. This was a surprising result. He tried other color combinations, but nothing came out the way he expected. Red, blue and yellow must not be the primary colors of the eye.

He began experimenting with other sets of three colors, and when he tried red, green and blue, something amazing happened. Spinning these three colors made them look very close to white. He then assumed that the eye must have red, green and blue receptors.

The next step was to make a chart of what you got when you mixed various proportions of each color. His paper disks could be adjusted so that more or less of each color could be shown. The percentages were carefully recorded on each side, so it was a very mathematical approach to describing color. (The resulting chart can be seen below.)

Then James thought of a possible problem— how could he be sure that his perception of color was the same as everyone else's? What if each person saw colors differently? Perhaps this triangle was just his own perception. So he asked many friends to come to his lab to try out the color wheel. What he found was that most people seemed to see colors exactly the way he did. After looking at the spinning wheel, they would pick out exactly the same colored paper sample that James had chosen. The only exception was people who were known to be color blind. Interestingly, not all color blind people would make the same mistake. There seemed to be different types of color blindness. James guessed that it was because people could have a problem with just one receptor. Most of his color blind volunteers seemed to be missing the red receptor and could not tell red from green.

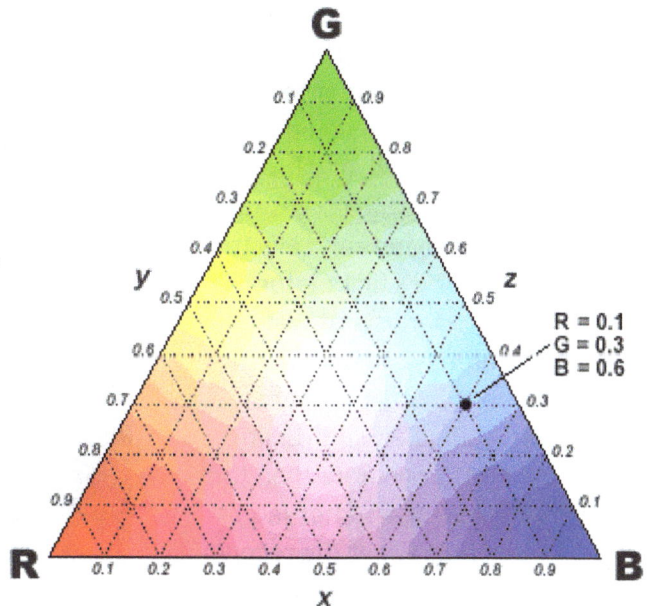

R = 0.1
G = 0.3
B = 0.6

Most people would stop there, satisfied that they had solved the mystery of color, but not James. He realized that he needed to experiment with light itself, not just colored paper. He used several prisms inside a wooden box to separate light into its spectrum colors. He put slits in the box so that he could let just one color come out of the box. Now he had a source of color directly from light. The box was fairly large, and took up most of the space on any table it sat on. (*2.6)

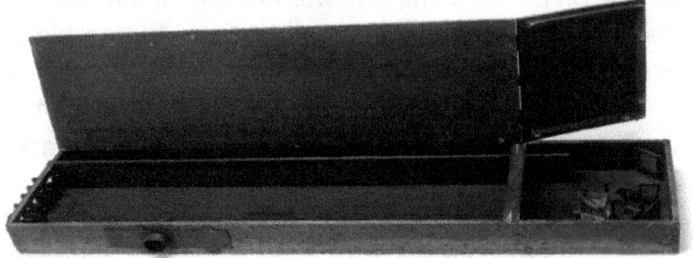

James would continue to experiment with the light box for many years. After he got married his wife would even help with the research. Meanwhile, James was also spending a few hours a day reading books about electricity and magnetism. He read about all the theories and experiments that had taken place in recent years. It seemed that scientists were divided in their views about magnetism. Some held that when magnets pull on a metal object, they are doing so "at a distance," meaning that nothing is happening in the space in between. This seemed to be the way gravity worked, so maybe magnetism was the same. The other view was that something was indeed happening in that empty space between magnet and metal. Michael Faraday was the main advocate for this theory. He claimed that the space all around a magnet was filled with invisible lines of force. His experiments had shown him that magnetism and electricity were linked, and he coined a new word: "electromagnetism." Then he used magnets and electricity to make the world's first motor.

Many scientists of that day scoffed at Faraday's ideas. Faraday had been born into a poor family and had never studied at a university. How could he know more than they did? Impossible! Years later James would help Faraday get the recognition he deserved.

Michael Faraday

As James read Faraday's writings, he followed along with each experiment, and imagined in his mind exactly what Faraday had seen. He marveled at Faraday's excellent experimental techniques. Faraday wrote about his failed experiments, not just his successes, and James admired that. James was convinced that Faraday was right. What Faraday lacked was the mathematical equations to back up his theory. Perhaps James could help, since he seemed to be good at using math to explain things.

This picture shows bits of iron that were sprinkled around a red magnet. The tiny pieces of iron are lined up along Faraday's invisible lines of force. Faraday had imagined these lines as "tentacles" wrapping around the outside, from pole to pole. James decided to think of them not as tentacles but as a steady stream of fluid and decided to use the word "flux" to describe them. The more dense the flux was at a certain point (at the poles, for example) the stronger the electric or magnetic force at that point. James guessed that perhaps this flux worked like the water in the streams at his home in Glenlair. In places where the stream was narrow, the water ran faster and

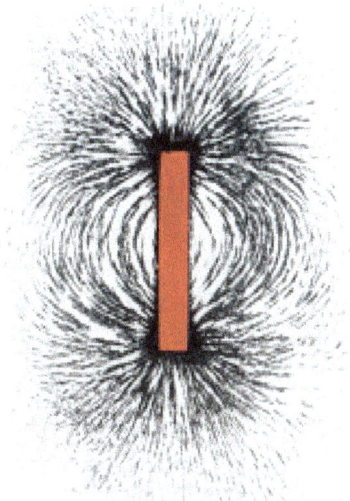

Tiny bits of iron can show the fields around a magnet.

harder. In places where the stream was wide, the current slowed down and was more gentle. What if electromagnetic flux worked like this?

James thought about other ways that fluids behave. Water flows from high areas, like mountains, to lower areas, like valleys. When it reach-

es the lowest point, such as a pond, it stops flowing and sits there. What makes fluids move is a difference in height or in pressure. Fluids always want to escape high pressure and move to areas of lower pressure.

The best thing about this fluid analogy was that there were already equations about fluids. He could start with those equations, and then somehow adapt them to his electromagnetic flux idea. He eventually came up with equations that seemed to describe exactly what Faraday had found in his experiments. He wrote a paper about his discoveries and called it: "On Faraday's Lines of Force." Michael Faraday was amazed when he received a copy of the paper. Someone had proven his theories!

Below you see the cover of his published paper, plus one of the illustrations. Notice that the word "science" is used. Previously, science had always been "natural philosophy." (*2.7)

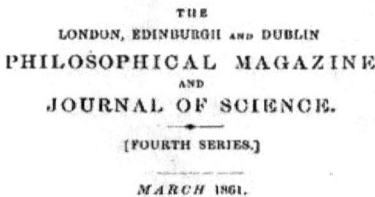

THE

LONDON, EDINBURGH AND DUBLIN

PHILOSOPHICAL MAGAZINE

AND

JOURNAL OF SCIENCE.

[FOURTH SERIES.]

MARCH 1861.

XXV. On Physical Lines of Force. By J. C. MAXWELL, Professor of Natural Philosophy in King's College, London*.

PART 1.—The Theory of Molecular Vortices applied to Magnetic Phenomena.

IN all phenomena involving attractions or repulsions, or any forces depending on the relative position of bodies, we have to determine the magnitude and direction of the force which would act on a given body, if placed in a given position.

In the case of a body acted on by the gravitation of a sphere, this force is inversely as the square of the distance, and in a straight line to the centre of the sphere. In the case of two attracting spheres, or of a body not spherical, the magnitude and direction of the force vary according to more complicated laws. In electric and magnetic phenomena, the magnitude and direction of the resultant force at any point is the main subject of investigation. Suppose that the direction of the force at any point is known, then, if we draw a line so that in every part of its course it coincides in direction with the force at that point, this line may be called a line of force, since it indicates the direction of the force in every part of its course.

By drawing a sufficient number of lines of force, we may indicate the direction of the force in every part of the space in which it acts.

Thus if we strew iron filings on paper near a magnet, each filing will be magnetized by induction, and the consecutive filings will unite by their opposite poles, so as to form fibres, and these fibres will indicate the direction of the lines of force. The beautiful illustration of the presence of magnetic force afforded by this experiment, naturally tends to make us think of

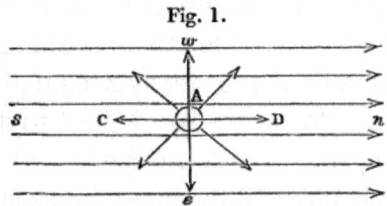

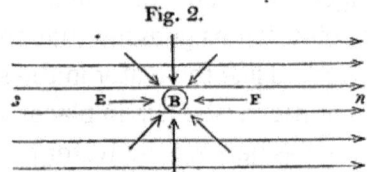

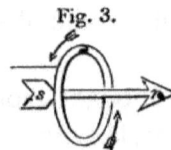

Fig. 1.

Fig. 2.

Fig. 3.

CHAPTER THREE

During this period of time when James was working on color theory and electromagnetism, he was also keeping up with his full time job as a teacher. Students loved his classes because he would go to a lot of trouble to set up interesting hands-on demonstrations. (Once when he was lecturing about how to calculate the rate at which water would escape from a hole, a pipe burst on campus and produced exactly the effect he was teaching about!) James wanted everyone to learn, not just the gifted and wealthy students that came to Cambridge. He helped to set up a Working Men's College that provided evening classes to ordinary folks who wanted more education. He didn't think he was above this kind of teaching and even did many lectures himself.

A Working Men's College in London

During school vacations (holidays) he would go home to Glenlair and help his father manage the house and farms. He did ordinary jobs like fixing machinery, paying bills, and caring for animals. During one of these vacations he was introduced to a younger cousin, Lizzie, and the two fell in love. The older relatives in the family said that although there wasn't a law saying that cousins couldn't marry, they thought it was unhealthy and would not allow James and Lizzie to continue their relationship. After a normal amount of heartbreak, both of them moved on with their lives and eventually married other people. It would not be too long until James would meet his future wife.

A cartoon from the 1800s, looking very much like James Maxwell (though not intentionally)

One day, James got a letter from Professor Forbes. He told James that there was a job opening at a small college (Marischal, shown on the right) in the city of Aberdeen, the largest city in the north of Scotland. The college was looking for a natural philosophy professor and they were willing to consider younger men, not just older ones. James's father wanted him to apply for the job because

Marischal College in Aberdeen, Scotland

Aberdeen was closer to Glenlair than Cambridge was. James did apply but did not get an answer right away. Not long after this, his father got very sick with a lung infection. Back in that day there were no antibiotics, and people often did not recover from infections. James left Cambridge and went home to take care of his father. He joked with his friends that he had turned into a nurse. One day a letter came, telling James that he could have the job at Aberdeen if he wanted it. His father was very happy and had a few weeks where he felt much better. But the infection came

back and one morning he died quite suddenly. James was now the master, or "laird," of Glenlair, a responsibility he took very seriously. He decided to take the job in Aberdeen, but not before he had made sure that all the farmers and workers at Glenlair knew how to run the estate by themselves when he was away.

James arrived in Aberdeen to take his job as a professor when he was 25 years old. The other professors were very nice to him and often invited him to dinner. The only thing he could complain about was that no one liked to tell jokes. They were all in their 50s and 60s and didn't need to tease and joke around like younger men did. James wrote this in a letter to a friend, "No jokes of any kind are understood here. I have not made one for two months, and if I feel one coming on I shall bite my tongue."

The other professors wanted to know what kind of teacher James would be, and they asked him to give an introductory speech about his teaching philosophy. He opened his speech by saying that he didn't want to simply teach science, but rather to use science to teach students to think for themselves. He also told everyone that experiments and demonstrations would be a key part of his teaching. One of his first lessons was on optics and he even brought in fish eyes and cow eyes for the students to dissect.

During that first year of teaching at Aberdeen, James heard about a contest that Cambridge University was sponsoring. They were giving a prize to anyone who could figure out what the rings of Saturn were made of. Astronomers had known about Saturn's rings for quite some time, but they did not know whether they were solid, liquid, or made of tiny particles. James began to think about Saturn's rings and how he might use math to solve the puzzle.

First, he read some math books and found some equations that he thought might be helpful. These were standard equations that everyone knew about, intended to be used to figure out problems involving water. James decided that the equations should work here, even without water. He found that if Saturn's rings were made of one solid piece and were very even all the way through, the constant rotation around the planet would eventually cause the ring to break apart. The only way a solid ring could be stable was if most of the mass was all clumped together in one place on the outside of the ring. Everyone knew this was not true because they could see in their telescopes that the ring was very thin all the way around and did not contain any thick places.

Next, he thought about what would happen if the rings were made of a fluid. He used equations by a famous mathematician named Fourier to show that the fluid would ripple about, and the waves would eventually cause the ring to break up into separate blobs. Again, astronomers could see Saturn well enough to know that the rings were not made of several large blobs.

Finally, he studied the possibility that the rings were made of many small particles. It was impossible to calculate what would happen to every particle, so he just imagined a small single ring made of equally spaced particles. The math equations showed that a ring like this would vibrate in four ways, but as long as the ring was not too dense in comparison with Saturn's density, it would be okay. Since the other two possibilities (solid and fluid) did not work at all, the rings must be made of many tiny particles.

In the 1980s, satellites Voyager 1 and 2 flew past Saturn and took pictures of the rings. They are, indeed, made of particles. This is a computer image showing what the particles might look like.

James wrote a paper about his findings and sent it to Cambridge. No one else had even entered the contest, so James was declared the winner of the prize. When physicists around the world read his paper and saw how he had used those math equations, they were very impressed. James wasn't just a number cruncher. His thinking showed great creativity. (*3.1)

During this time, the other professors continued to invite James into their homes for dinner, but the home he visited most often was that of the principle of the college. It just so happened that the principal had an unmarried daughter 7 years older than James, and the two of them got along very well together. Perhaps, at first, it reminded James of spending time with his cousin, Jemima, who was 8 years older than him. Eventually, James and Kathryn decided that they wanted to spend the rest of their lives together. James wrote a poem for Kathryn, asking her to join him at Glenlair. Here are the first lines from the poem:

> Will you come along with me, in the fresh spring tide,
> My comforter to be, through the world so wide?
> Will you come and learn the ways, a student spends his days,
> On the bonny bonny braes, of our ain burnside?

"Ain" means "own," "bonny" means "beautiful," "braes" means "hillside," and "burnside" refers to Glenlair.

In June of 1858 they were married, with James's best friend from childhood, Lewis Campbell, as their best man. Kathryn would turn out to be a very good companion, indeed, and would help James with many of his experiments. This picture shows them with their dog, Toby. From that time on, James was never without a dog named Toby at his side. When the current Toby died, he would get a new Toby. Sometimes he would even take Toby with him to work.

Kathryn shared James's faith in God, and they started every day in their household with Bible reading and prayer.

James had not yet published anything about his work with red, green and blue light. Now was the time to finish that up so he could move on to other things. A craftsman in Aberdeen made him an improved light box that gave better control of the proportions of colors, and Kathryn helped to make all the final observations. The research paper was written and published, and then James felt free to explore other topics. He would eventually get back to the topic that would make him famous, electricity and magnetism, but first something else caught his attention.

In 1859 he read a paper by a German physicist who was puzzling over how gases diffuse into air. For example, if a perfume bottle is opened in one corner of a room, the smell will gradually travel farther and farther from the bottle until eventually the whole room is filled with the smell. Scientists were already pretty sure that gases consisted of very small particles, and that these particles likely collided with each other as

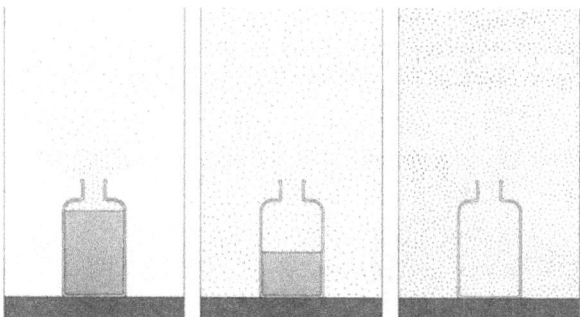

they went through the air. These collisions between the particles would help to spread them out. What puzzled this German scientist was that the math equations he was using predicted that the gas molecules would have to travel at an enormously high speed. If they were traveling this fast, the smell of a gas should spread much faster than it actually did in a real situation. What was going on? James became fascinated with this problem and decided to see if he could solve it.

When he was a student at Edinburgh, James had been taught that the pressure of gases was caused by the repulsion of the particles (which we now call molecules). This would be a bit like the repulsion that two north poles of a magnet have for each other. The gas theory stated that the gas particles repelled each other for some unknown reason. To James, this didn't make sense. He preferred another theory of gases, the kinetic theory. ("Kinetic" means "moving.") This theory stated that gas particles are in constant motion. If the gas is heated, they speed up, and if they are cooled, they slow down. As the particles move, they bump into each other. This can change the direction in which they are traveling, but it's a random process with particles having an equal chance of going in any direction. James could not explain in words why he thought the kinetic theory was right. He just knew it was. This is called "intuition" and James had a lot of it.

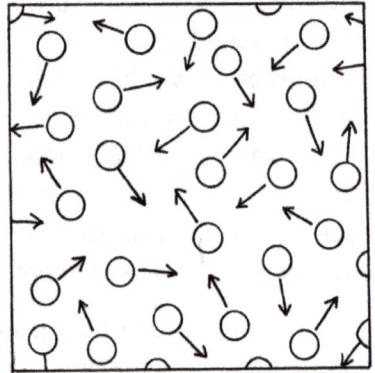

Once again, the solution to the problem was to be creative. Everyone else had tried to solve the problem by thinking about all the individual particles. James thought it might be okay to simply consider the volume of gas as a whole, and think about what this entire volume might do. Again, he used math equations that were already in existence, but applied them in a way no one had thought of. He borrowed equations from a branch of math called "statistics." This is the branch of math that tells you how likely something is to occur. For the first time ever, James created a set of statistical equations that applied to gases. Once he had done it, everyone thought to themselves, "Now why didn't I think of that?" Other physicists and chemists picked up this idea and ran with it, and today we have a whole branch of science called statistical mechanics.

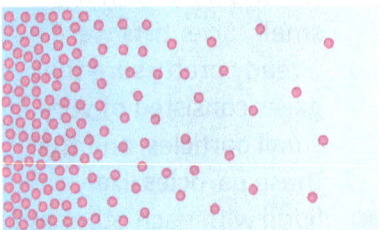

James was extremely clever, but he was not perfect. As he continued to work on more theories of gases, he began to make mistakes. He made an incorrect prediction about what would happen to the thickness of gases as temperature increased. Experiments proved him wrong. He also made calculation errors in his equations about the nature of the energy in gases, mainly because he forgot to convert kilograms to pounds, and hours to seconds. (Have you ever done that?)

James went on to other things eventually, and his ideas about gases were taken up by another man, Ludwig Boltzmann from Austria. Boltzmann spent most of his life studying this kind of science, but some of the discoveries bear both men's names, most famously the "Maxwell-Boltzmann distribution of molecular energies." (*3.2)

Ludwig Boltzmann

As usual, James was doing all this research in his spare time. His full-time job was still teaching classes. We have quite a few firsthand accounts of what James was like as a teacher. His brain was always full of pictures of what he was thinking, so he used a lot of picture analogies. (For instance, electricity is like water.) This was okay to a point, but sometimes he would switch to different picture images so many times in one lecture that the students found it confusing rather than helpful. James would also make arithmetic mistakes on the blackboard and, if the math was very complicated, end up wasting lecture time trying to track down his mistake. What the students liked best about his teaching was his enthusiasm and the way he cared about his students. He would often spend hours talking with small groups after class, and would use his own library privileges to check out books for them.

James also continued doing evening lectures at the local Working Men's College. One man who went to those lectures later wrote about them. He particularly remembered the time that Prof. Maxwell brought a machine to class that would generate static electricity. He would ask for a volunteer to stand on an insulating mat, then apply so much static to them that all their hairs stood on end. Today, this is not an uncommon site at a hands-on science center, but back then this was quite an exotic experience! (*3.3)

Big changes were about to take place in Aberdeen. The colleges had decided to combine to make one large university. This meant that some professors would lose their jobs. The University of Aberdeen would need only one head of the natural philosophy (science) department, and James was by far the youngest and least experienced person applying for the job. It was not a surprise when James was told he had to leave Aberdeen and look for a teaching job at another university. Professor Forbes was retiring, so James applied to be his successor. No luck. His old school mate, Peter Tait got the job. Then another opportunity came up at King's College in London, and this time James was selected.

James returned to Glenlair for the summer, knowing that in the fall he would be moving to London. He wanted to make the summer special for Kathryn, so they went to a horse fair and bought her a bay pony. Unfortunately, the pony was not the only thing they brought back from the fair. They were exposed to the viral disease called small pox, and James came down with it. Small pox was a deadly disease and has killed thousands upon thousands of people in past centuries. In the 1900s, a vaccine was developed to prevent small pox infections, but in the 1800s small pox was

still a major threat. James became very ill for several weeks as his body struggled to fight the virus. Kathryn was at his side doing everything she could to help him through it. Eventually, his body won the battle and he began to recover. Once he was well enough to be outside again, he devoted many days to training the pony, which they had named Charlie.

James and Kathryn took Charlie (and Toby) with them when they moved to London. They rented a house in the neighborhood called Kensington. Nowadays, Kensington doesn't have horse stables, but back then they were able to rent a stable for Charlie so Kathryn would be able to go out riding any day that the weather was nice. James often rented a horse for the afternoon so they could ride together. This photo shows how their London house looks today.

King's College, London, as it looks today. The object in the middle of the skyline is a Ferris wheel called the London Eye. None of these tall structures were there in Maxwell's day.

James was expected to introduce himself to the college with an inaugural lecture. At age 29, he had already given several of these, so he used the same themes that he had before, stressing the importance of helping students to think for themselves. He said, "I know the tendency of the human mind is to do anything rather than think. But without understanding the principles on which formulas depend, the formulas themselves are mere rubbish." He also wanted to include his own opinion about scientific discovery. At this point in time, some were beginning to complain that pretty much everything about science had been discovered, and there was very little left for newcomers to do. James said, "The present generation has no right to complain about the great discoveries already made, as if they leave no room for further enterprise. These discoveries have only given science a wider boundary."

Living in London gave James a new opportunity he had never had before—attending lectures at the Royal Institution, the very place where Michael Faraday had presented his lectures about electromagnetism, as shown here.

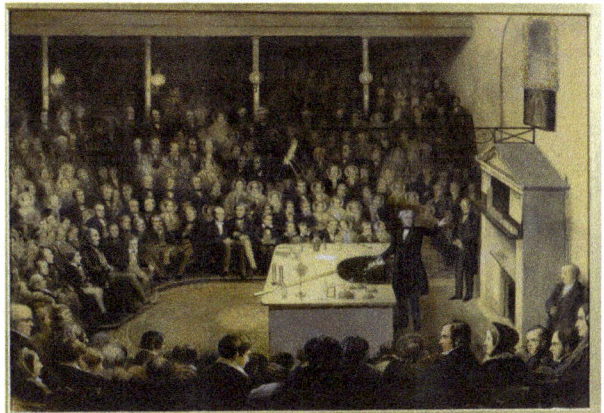

Michael Faraday was elderly by now and rarely gave lectures, but he did attend some, and it was at one of these lectures he attended that he was introduced to James Maxwell. They had corresponded by letter for several years, and were very excited to finally meet each other in person. This meeting would give James renewed interest in electromagnetism and get him back to researching this topic. But first, he had return to the topic of color vision one last time because the Royal Institution had asked him to give a lecture about it.

Michael Faraday (on right) and another scientist demonstrate electrochemistry.

James wanted the lecture to start with a dramatic demonstration showing how red, green and blue light can combine to make any color. He could not demonstrate his light box because it could only be used by one person at a time. He had a spinning top, but that was still no good because people at the back of the auditorium would never be able to see it. What he needed was something large and colorful that everyone could see. He hit upon the perfect idea—he would make the world's first color photograph and project it onto a screen for everyone to see. A friend of his was an expert at making photographs, which at this time was a complex process involving glass or metal plates that had to be developed using chemicals. But these photographs would only be black and white. Where would the color come from?

James's idea was to take three separate photographs of the same object, but put a colored filter over the lens each time. The photographs would still be black and white, but if they were all projected onto a screen at once, and if those same color filters were put in front of the corresponding projectors, the result should be to produce a color picture of the object.

Now here is where luck stepped in. James had a tartan ribbon that he thought might look nice in color. What he didn't know was that this ribbon had some unique features that would allow his idea to work, even though his theory was not perfect. He was indeed able to produce a color image of the ribbon using the technique he had planned. The actual image is shown at the top of the next page. The audience that night at the Royal Society was stunned. No one had ever seen a colored photograph.

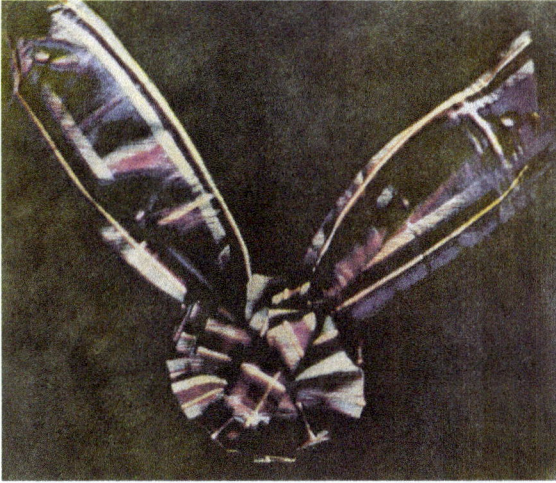

Immediately, other scientists tried this new technique, but no one could get it to work. Photographic plates seemed not to be sensitive to red light. The red part of an image would always be missing. How had James gotten those pink colors in his ribbon?

Photographers puzzled over this for more than 100 years, trying to figure out how he had done it. Finally, some researchers at Kodak solved the mystery. The makers of that particular ribbon had used a red dye that also contained a chemical that gave off ultraviolet light. It was the UV light that the plate had picked up, not the red. Not only that, but one of the chemicals used to develop the picture just happened to be sensitive to UV light. If this had not been true, the plate would not have shown the red image. It was an amazing coincidence. The result was so impressive that James was immediately elected as a member of the Royal Society. (This was an exclusive group of the best scientists in England.) [*3.4]

Finally, James could go back to thinking about electricity and magnetism. Would he be able to remember what he was thinking five years ago when he wrote his first paper on this topic? Could he pick up where he left off and continue on? James was a believer in subconscious thought, the idea that our brains go on thinking about things at a very deep level even when we are not actively thinking about them. He wrote a short poem about this:

There are powers and thoughts within us, that we know not till they rise,
Through the stream of conscious action from where Self in secret lies.
But where will and sense are silent, by the thoughts that come and go,
We may trace the rocks and eddies in the hidden depths below.

James trusted that his brain would not only remember, but would come up with new ideas he had not thought of before. (To help him think, he would take Toby with him to work every day, and talk out loud as if Toby could understand.)

First, he clarified the problem. He needed to come up with some equations that could explain the following observable facts:

1) "Like" forces repel and "unlike" forces attract. These forces follow the "inverse square law" meaning that as the distance increases arithmetically (1, 2, 3, 4, 5...) the force decreases geometrically (1, 1/4, 1/9, 1/16. 1/25...)

2) North and south poles always occur in pairs on a magnet. You never have north without south.

3) An electrical current in a wire creates a circular magnetic field around the wire.

4) A changing magnetic field produces an electric current.

James decided that he needed a better analogy. The analogy he had used previously, imagining electricity to be like water, had been a good place to start, but now it felt like a dead end. He needed something else. He began thinking about magnetic fields and how they go out at right angles to an electric current. How could this be represented? What if he imagined the space around the wire, the area of the magnetic field, to be filled with tiny rotating balls? He wasn't thinking about atoms. No, these were completely imaginary balls. He thought of them as "spinning cells." All they needed to do was explain why the magnetic field was at a right angle to the wire, not parallel to it. (In this case, the word "cell" is not referring to anything living, just an individual spinning unit.)

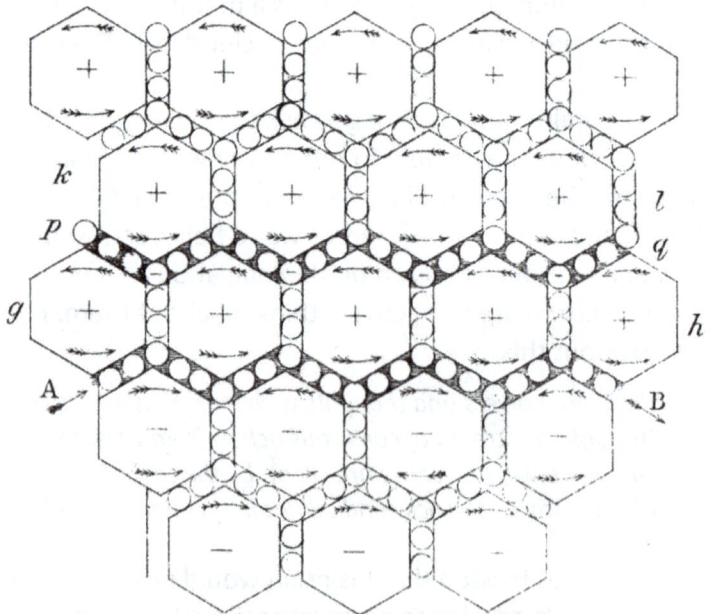

An illustration from his finished publication about this topic. This diagram shows his "spinning cells."

He imagined that if the cells were all rotating in the same direction, they would exert a combined pushing force on anything in that space. The axes of the cells would be parallel to the wire. Perhaps these little cells bulged out a bit at their equators, like the Earth did? That bulge would be where the cells were actually touching and pushing on each other. If the balls were rotating in opposite directions that would cause problems, so he decided to imagine even smaller balls between them. Maybe he could even imagine these smaller "wheels" to be particles of electricity, and they exerted the force that set the larger balls spinning in one direction. He didn't need to explain it any more than that. He just needed the rough idea to work.

Furthermore, if he imagined that the lines of tiny balls were single lines of electrical current, he could extend them to wires outside of his spinning cell area, and by controlling the electrical particles, control the magnetic force created by the spinning balls.

If all of this sounds confusing, don't worry about it. The main point is to understand that James was cleverly drawing upon his knowledge of how real gears and balls work and applying this knowledge to an unknown situation. James went on to continue the analogy and found that it worked for all four of those observable facts listed on the previous page. (If you are intrigued and want to read more about the spinning cells in great detail, you can find this information in a book called "The Man Who Changed Everything" by Basil Mahon.)

The most exciting thing about this analogy from James's point of view was that he might be able to use it to make predictions about things yet to be discovered. For example, according to his spinning wheels model, electricity and magnetism are permanently bound together and cannot be separated. They are the same thing, like two sides of one coin. Also, it seemed to him that there could be a tiny "twitch" in the system if something started to push on the wheels but then they were stopped by the "friction" between them.

Maxwell during his early years at King's College

37

Would this idea correlate to an observed phenomenon some day? Further, the energy movement in these spinning cells would act more like wave movement than anything else. Would electromagnetic waves behave like water or sound waves? His model predicted that this would be so. These would be questions for other scientists to investigate.

Right in the middle of all this work on the theory, the school term ended and James had to go home to Glenlair to take care of the estate. Oddly enough, he didn't take any math or science books home with him. He planned on spending the summer mending roofs and stone walls, taking walks along the creek and watching the northern lights on long summer evenings. As he was doing all this work, however, his brain kept thinking about the theory, and by the time he returned to London in the fall, he was ready to write up his theory and publish it. (*3.5)

a creek in Scotland

CHAPTER FOUR

James returned to King's College in the fall, not just ready to begin a new semester of teaching, but ready to write and publish a new paper about electricity and magnetism. In this paper he introduced to the world his analogy of spinning cells and to the idea that light was a type of electromagnetic wave. It was a brilliant paper, but James was not entirely satisfied. His goal was to eventually understand electromagnetism well enough that he did not need the spinning cells analogy anymore. It would be two more years before he would reach this goal.

Meanwhile, James and Katherine continued to do more work with the light box. Every guest that came to their home was asked to look into the light box and describe the colors they saw. It was a tedious process, requiring many adjustments to the lenses and prisms, but eventually they collected enough information to be able to describe the difference in color vision between normal people and those who are color blind.

James also wanted to follow up on his theory about gases and run some experiments to see if his mathematical predictions were

James and Katherine (with Toby, of course)

accurate. According to his equations, gases under pressure would not experience any change in their viscosity *(viss-KOSS-i-tee)*. Viscosity is a measure of how thick or how runny a substance is. For example, molasses and ketchup are very viscous *(viss-kuss)*. They are thick and run very slowly. Water and milk pour easily because they have a low viscosity. Honey is more viscous than water but not as viscous as molasses. We don't think of gases as being viscous at all, but they do have a small degree of viscosity. The problem was how to measure it. James came up with an ingenious invention to do just that.

The choice of which gas to use was obvious: regular air was plentiful and non-toxic. The air would need to be enclosed in a glass case that could be pressurized. To measure the viscosity, James used a type of pendulum that rotated around its axis, like a top. There were little "fins" mounted on the round, flat, plate which would catch the air like the sail on a boat. If the air became more viscous as it was pressurized, it would slow down the rotation of the pendulum. It was a brilliant idea, but ended up being more difficult than he had expected. A number of things went wrong with the equipment. The glass case not only leaked but eventually exploded. James did not give up, however, and was ultimately able to get some good data. The experiments seemed to prove that James's equations were correct. Gases do not become more viscous as their pressure increases.

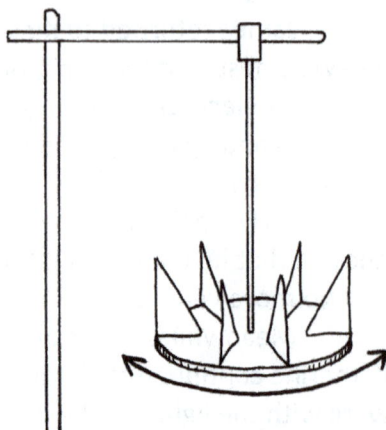

On a torsional pendulum, the "bob" (weight) at the bottom spins instead of swinging back and forth.

Next, James wanted to see if this held true for increased temperature. He predicted that he would get the same results, and that increased temperature would not produce increased viscosity. For this experiment, the glass case would have to be wrapped in a metal lining that could be filled with hot water or with hot steam. Katherine was a great help in this experiment, carrying both boiling water and ice up to the attic. She wrapped the entire case in some old blankets to keep it insulated, and even put a feather cushion over the top. It's likely that they both watched the pendulum rotate, and made a precise count of how many rotations it made before coming to a stop. If the air was more viscous it would slow the pendulum down and cause it to come to a stop more quickly. They collected a lot of data, and when James analyzed it he discovered that his prediction was absolutely wrong! Gases <u>did</u> become

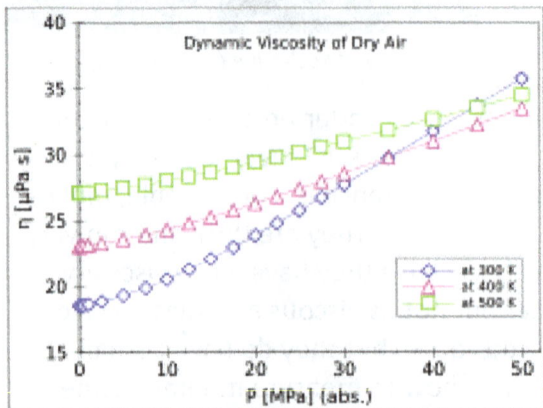

James's data might have looked something like this.

more viscous as their temperature increased. James would have to do some re-thinking of his original theory. He did not have any immediate ideas, so he decided to let the question rest for a while and allow his subconscious mind to work on the problem while he was busy doing other things. [*4.1]

James was then obliged to turn his attention to something that was not glamorous at all. In fact, many people would say it was downright boring. Scientists needed accurate ways to measure these new discoveries about electricity and magnetism. How can electromagnetic force be measured as it goes through a wire? James teamed up with some other scientists to create a complicated magnetic device that would hopefully measure the resistance of a wire to the electricity passing through it. Electrical resistance is a fundamental property of materials and would become an important property to measure. Their measuring device was an ingenious contraption that involved a spinning coil of wire that was sensitive to the earth's magnetic field. It needed to be very sensitive to give

them the readings they needed, but they found that it was so sensitive that it could detect iron barges going up and down the Thames River nearby. They could only get accurate readings when no boats or barges were in sight.

Iron steam boats like this were often seen on the Thames during the years that James was at King's College, London.

The last part of the experiment involved a precise measurement of the length of the wire that they had wound into a coil. The unwound wire did not want to lie straight, so they had to take it to a university building that had wooden floors with wide cracks between the boards. They pushed the wire down into the long, straight cracks and were able to get the measurement they needed.

They decided to create a new unit of measurement called the Ohm, named after Georg Ohm, who had come up with mathematical laws to describe how electrical current flowed through wires. Ohms tell how resistant a material is to having electricity flow through it. If you want to reduce the flow of current coming into an appliance, you can use a resistor. [*4.2]

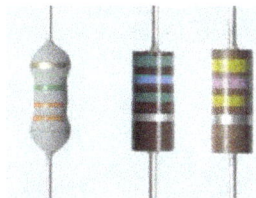

The bands of color on these resistors are codes that give information about the resistor, including how many Ohms it is.

41

As part of his teaching duties, James was required to keep up with all the latest technological developments in the world of physics and engineering. He read about the work of a Scottish civil engineer who had come up with a better way to build trusses (supports) on bridges. Some of his students were likely going to become civil engineers, so this was important information for them to know. However, James realized that there was a simpler way to do the math. Basically, he showed that you can use simple geometric drawings to represent a real structure. He quickly wrote up a paper about it, called "On Reciprocal Figures and Diagrams of Force." This led to several more papers, as his mind quickly jumped from one idea to the next. Today, these methods are common practice in the field of civil engineering, although computers have made the work much easier.

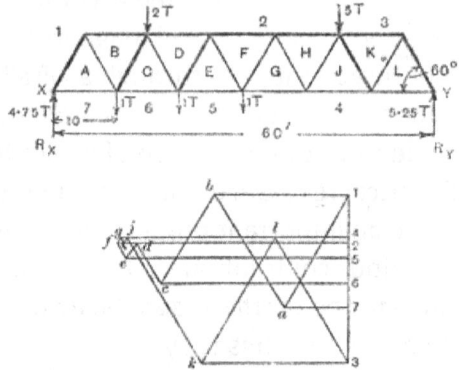

Finally, after this work was done, James was able to turn his attention back to his thoughts about the relationship between electricity and magnetism. Had his brain been working on the problem subconsciously all this time? Indeed it had, and in seemingly no time, he was ready (in 1864) to write a paper that would rock the scientific world. In a letter to his cousin, he expressed his feelings about his soon-to-be-published paper. "I also have a paper afloat, with an electromagnetic theory of light, which, till I am convinced to the contrary, I hold to be great guns." Great guns was an understatement. James knew he was on the brink of one of the major scientific discoveries of all time. The equations that described his theory were shockingly simple (as far as physics equations go). This is what they look like:

$$\nabla \cdot \mathbf{E} = \frac{\rho}{\varepsilon_0}$$

$$\nabla \cdot \mathbf{B} = 0$$

$$\nabla \times \mathbf{E} = -\frac{\partial \mathbf{B}}{\partial t}$$

$$\nabla \times \mathbf{B} = \mu_0 \left(\mathbf{J} + \varepsilon_0 \frac{\partial \mathbf{E}}{\partial t} \right)$$

These are known as "Maxwell's Equations," though each equation bears the name of a previous scientist who had already worked on electricity and magnetism. (Gauss's Law of Electricity, Gauss's Law of Magnetism, Faraday's Law of Induction, and Ampere's Law)

NOTE: The equations don't always look like this. If you do an image search online, you will find that there are many ways of writing these equations. The meaning, however, is always the same.

Here is another version of these equations (just one of the many variations). "E" stands for electrical force and "H" stands for magnetic force. They are written in bold (dark) letters to show that they are "vectors" which means they have direction as well as strength. The letters "div" stand for "divergence, and tells whether the force is directed outward or inward. The word "curl" means

1) div $\mathbf{E} = 0$

2) div $\mathbf{H} = 0$

3) curl $\mathbf{E} = -(1/c)\ d\mathbf{H}/dt$

4) curl $\mathbf{H} = (1/c)\ d\mathbf{E}/dt$

exactly that—something curls around something else. The terms "dH/dt" and "dE/dt" are rates of change over time. The letter "c" represents a number that James determined was necessary in order to make the equations work. In James's mind, this letter "c" was like the tiny spheres in between the larger spheres in his spinning cell analogy. This is the same letter "c" that would appear in Einstein's famous "e=mc^2" equation several decades later. "C" has a numerical value: 300,000 kilometers per second (186,282 miles per second). Most of us know "c" as the speed of light, but it is also the speed of any type of electromagnetic radiation including radio waves, x-rays, infrared rays, and microwaves.

So what do these equations mean? The first one says that there is no outward or inward electrical force around any single point. The second one says the same thing about magnetic force, and also implies that magnets will always have a pair of poles, north and south. This is why if you break a magnet in half, both halves will then have north and south poles. Break those in half and your will get four magnets with north and south poles.

The third equation says that when the magnetic force changes, it wraps a circular electrical field around itself. That minus sign means that the direction of the curling is counterclockwise. The fourth equation says that an electric current will curl a magnetic field around itself in a clockwise direction as it travels. This explains the behavior of a compass near an electrical field. If you switch the direction that the electricity is traveling, the compass will spin around and point the other way, but still at a right angle to the wire. The third and fourth equations together tell us that electrical and magnetic waves always travel together and can't be separated. These waves are transverse waves, like the

"up and down" waves we see in the ocean. However, in an electromagnetic wave, the electricity and the magnetism don't go up and down together, but at right angles to each other. Again, this helps to explain the behavior of a compass needle in the presence of an electrical wire.

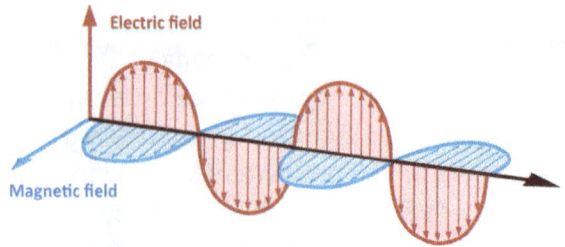

Today we recognize that these equations are foundational for just about every electronic project or device. They brilliantly unify previous ideas about electricity, magnetism and light. We hail these equations as being some of the most important science equations of all time. But how was the paper received in 1864? Was it immediately embraced by physicists everywhere? Hardly. It was more like stunned silence. Maxwell's equations indicated that electromagnetic forces, fields, and waves where everywhere. If these waves really did exist, there would have to be many types—an array of different wavelengths. So far, no scientist had ever claimed to have detected any of these waves (other than visible light). It would be twenty years before anyone would claim to have discovered more forms of electromagnetic waves. During those twenty years, Maxwell's equations would remain just an interesting theory. (*4.3)

At the height of his career, James decided to leave King's College. We don't know exactly why, but it seems he felt a strong responsibility to his estate at Glenlair. He was the "laird" of the estate and it was hard to manage the estate while living in London. He resigned from his professorship and moved back home.

Back in Glenlair, James threw all his energies into upgrading many features of the estate, including a large addition to the house. His father had drawn up plans for this addition but had never had the funds to complete them, so James was able to bring his father's vision into reality. He also helped with building projects in his local community. He donated a substantial amount of money to the village church so they could build a house that could be occupied by whoever was the current minister.

James and Katherine loved children, and the only great sadness in their life was that they were unable to have children of their own. James had always said that people should not sit around thinking about what might have been but get busy and make the most of what they have. So he and Katherine showered the local children with love and affection. James loved to amuse them with games and interesting tricks. When they saw that the local school building was small and not in good repair, they donated part of Glenlair's land for a larger and better school building.

While Glenlair was being renovated, James and Katherine decided to get out of the way of the workers and go on a long vacation. Neither of them had ever traveled much, so they decided to do a once-in-a-lifetime trip to Italy to see all the famous historical sites. The trip had some unexpected adventure added to it—the ship had to be quarantined when they got to Marseilles, France. That meant that no one was allowed to get on or off the ship for many days. This caused great distress among the passengers, and James volunteered to be in charge of the distribution of water on board. He spent many days hauling buckets of fresh water to passengers. When they finally got to Florence, Italy, who should they run into but James's best friend from childhood, Lewis Campbell! What an amazing coincidence! James and Katherine must have had a wonderful time visiting with Lewis and his wife.

James and Katherine took Italian language lessons while in Italy. James found it similar enough to French and German that he was able to catch on quickly. He was soon able to talk about scientific matters in Italian with some Italian physicists in the city of Pisa. Pisa is the city with the famous Leaning Tower, the building from which Galileo was said to have dropped the balls of different weights, proving that Aristotle was wrong when he said that heavy objects fall faster.

After they got back to Glenlair, James spent most of his time writing and re-writing various scientific papers. Publishing is a time-consuming process and James had to write all his papers himself since no one else understood what he was thinking. He also spent a lot of time writing letters to his colleagues at Cambridge, and to his friends Peter (P. G.) Tait and William Thomson ("Lord Kelvin"). James continued to be a member of various scientific and mathematical societies and traveled to London occasionally to take part in their annual meetings.

James thought that a new textbook needed to be written about the science of heat, so he started writing one. He intended to simply write a useful textbook, but he ended up re-thinking the topic as he wrote. His publication, "The Theory of Heat," introduced new mathematical relationships between temperature, volume and pressure that are now known as "Maxwell's Relations." He used what is known as a "thought experiment" to discuss the idea of heat flow. (see page 76) Scientists had determined that heat always flows in one direction: from hot to cold. The flow of heat continues until "equilibrium" is reached, meaning that both the hot and cold areas are now the same temperature. This is what happens to a hot cup of coffee that sits on the table for a few hours. The coffee will eventually become the same temperature as the air around it. We all know that this situation will never reverse itself naturally. The coffee will never again be hot unless some energy is put back into it. (*4.4)

Other things caught James's attention, too. He was outside a lot, walking across the Scottish countryside. He looked at all the hills and "dales" (valleys) and realized that there must be a relationship between them. He saw that the number of hills must always be one more than the

number of dales. He came up with equations about this topic and published a paper called "On Hills and Dales." This branch of mathematics would eventually be called "topology."

What would turn out to be the final chapter in James's career began in 1871. Cambridge University felt like it was lagging far behind other universities when it came to scientific research. Many other universities, both in Britain and in Europe, were doing ground-breaking experiments. Several financial donors volunteered to give Cambridge the money to build a state-of-the-art laboratory, which would be named after a brilliant (but rather odd) scientist named Henry Cavendish. (Henry's nephew had been one of James's college friends.)

Cavendish Laboratory would be looking for a director. Cambridge's first choice was James's friend William Thomson, who was working in his own lab in Glasgow, Scotland. Thomson turned them down. Second choice was Hermann Helmholtz of Germany, but he also declined. Third was James Clerk Maxwell, who took a long time to think about it, but eventually said he would take the job, but only if he could quit after one year, in case it didn't work out so well.

James started his work by visiting all the best labs in Britain. He learned many tips on how to construct a building that was good for lab work. For example, it would need a 50-foot tower with a huge water basin at the top, so that the water pressure could be used to run a strong vacuum pump.

The original Cavendish Laboratory building

James was required to give a lecture to celebrate the official opening of the lab, and in that lecture he stated his belief that science and math should be taught in various ways because people's brains have different learning styles. He said that scientific truths should be presented in various forms, so that everyone has a chance to learn. There should be not only books, but also colorful pictures and live demonstrations. Science should be for every type of learner, offering experiences in many formats: visual, verbal, and kinetic (doing). This was not the general attitude of educators in the 1800s. James's ideas were ahead of his time. In the 21st century, we are so used to hands-on science that we forget how recent this idea is. Until the 1970s, museums were "looking-only" and classrooms were almost always "sit-and-listen."

A funny incident occurred at about this time. Not everyone believed that experiments were important in education, and they doubted that Cavendish Lab would be successful. One staunchly anti-experiment professor told James that the only evidence a student needed was the word of his teacher. One day James happened to see this man walking past the lab, and he invited him to come in and see a demonstration of the conical refraction of light. The man replied, "No! I have been teaching it all my life and don't want my ideas upset by actually seeing it!"

Not too long after the lab had opened, James's friend, the Duke of Cavendish, (who had provided a lot of the money to build the lab), brought James a huge stack of papers. The Duke had found hundreds of pages of scientific notes that his uncle, Henry Cavendish, had written many years ago. These papers had only recently been discovered because Henry had never published them. The Duke's Uncle Henry had been a very strange person indeed. (Modern psychologists think that Henry undoubtedly suffered from a fairly severe case of autism.) Henry disliked any and all interactions with other humans, especially women. He told his servants he never wanted to see them and that they should communicate with him by sliding written notes under his door. Only on rare occasions would he leave the house, and it was always to go to a scientific meeting.

This is the only picture we have of Henry Cavendish. He refused to have his picture taken or his portrait painted. This sketch was done by an artist who remembered him.

Publishing his papers would have required more contact with people, so Henry never did anything with the papers.

Upon reading through some of these papers, the Duke realized that his uncle had made many key discoveries about electricity years before anyone else. If his uncle had published his papers, what we now called Ohm's Law would have been called Cavendish's Law. Henry discovered Coulomb's Law decades before Coulomb did. The Duke was looking for someone to read through all these papers and possibly get them published so that his uncle's work could get the credit it deserved. James was an expert in the science of electricity and he also knew how to write and publish, so he seemed the perfect person for this job. James had plenty of his own work to write about, but being a generous and kind person, he agreed to read Henry's papers.

James found that Henry had indeed been the first person to make many scientific discoveries. Henry had been much like James, being good at working with equipment, often finding clever ways of use simple things to test his ideas. For example, Henry thought that electricity followed the "inverse square law" first suggested by Isaac Newton to describe the gravitational attraction between stars and planets. The inverse square law says that when you move twice as far away, the strength of the force is cut by 2^2, or 4. If you move three times as far away, the force is reduced

THE INVERSE SQUARE RULE

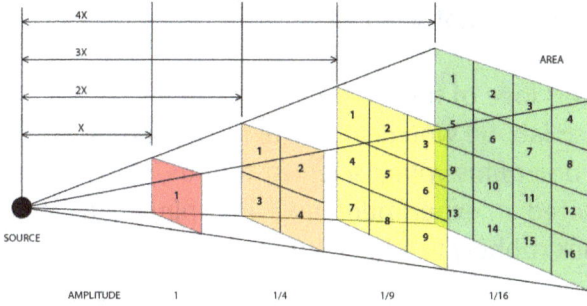

by 3^2, or 9. Four times as far would give you 4^2, or 16. Henry believed that electricity would follow this pattern. So he rigged up a shocking device, and gave himself electrical shocks of various current strengths and noticed how far up his arm the electricity traveled. James was intrigued by this experiment and decided to try it himself. He also tried it on many students and found that people who have tough, calloused hands have greater electrical resistance. A visiting scientist from America was appalled to find the famous James Clerk Maxwell playing with electricity and giving himself shocks!

James also got involved in writing and editing for the Encyclopedia Britannica, which had begun in 1768 and was then putting out its 9th edition. James wrote many articles for the encyclopedia and tried to make sure they would be understandable to the average reader. He also wrote articles for a new journal called "Nature," which would become one of the most prestigious science journals of the 20th century. (It continues to be prestigious today. If your article gets published in "Nature" you gain international recognition.)

James was no longer young. He was in his 40s and students looked up to him as a fatherly figure. He had a passion for inspiring and encouraging his students. Though he could spot talented students and would make sure they were given the opportunities they needed, he didn't neglect his other students and believed that everyone had potential and had a role to play in the unfolding history of the world. He never spoke harshly or criticized anyone, though he would never back down from intellectual

arguments when he believed he was right. Under Maxwell's leadership, Cavendish lab flourished. It quickly become an exciting place to work and began producing excellent experimental scientists.

James never gave up his love of funny poetry. Even in his last years, as director of Cavendish lab, he would still write funny poems about various events. When one of his young scientist friends, Henry Rowland, successfully completed an experiment involving spinning a disc made of ebonite (a hard, black rubber) in a beam of light, James wrote:

> The mounted disc of ebonite
> Has whirled before, nor whirled in vain;
> Rowland of Troy, that doughty knight,
> Convection currents did obtain
> In such a disc, of power to wheedle,
> From its loved north, the subtle needle.

During the 1870s, many famous inventions were being developed. Alexander Graham Bell had just invented the telephone, and Thomas Edison was working on the phonograph. There were hundreds of other inventions that are now less famous than those of Bell and Edison, including a little curiosity called the radiometer. William Crookes, the inventor of the radiometer, could not even explain how it worked. It was a glass bulb containing something that looked like a weather vane. The vanes were painted white on one side and black on the other. The glass bulb had most of the air pumped out, but it was not a perfect vacuum. When the radiometer was set in the sun, the vanes began to spin around. Physicists everywhere started making their own radiometers so they could study them, but no one could come up with a plausible explanation of why the vanes were moving.

James took a shot at explaining the mystery of the radiometer. He understood what sunlight was (waves of electromagnetic energy) and he had a lot of experience working with gases. There were still a few gas molecules left inside the radiometer, and those molecules would follow the laws of how gases behaved. He thought that the black side of the of vanes was heating up because of the sun and this was causing the air molecules that were close by to speed up. These fast-moving molecules would then create a very small flow of air around and past the edges of the vanes. He called this the "slip current." Some of the molecules in this slip current would hit the edge of the vane as they passed, nudging the vane a bit in that direction. After thousands or millions of these nudges, the vane would begin moving around. He published a paper about this theory and even came up with some new equations about "rarefied" gases (when there are very few molecules present). This paper, along with all the interest in radiometers, created a new branch of science, the study of rarefied gases. The place where this is most important is the study of the upper atmosphere.

The puzzle of the radiometer was not finally solved until the 1920s. It turned out that Maxwell was mostly wrong, although his idea of slip currents was basically valid. It was determined that it was the force of the heated air molecules hitting the black side of the vane that was driving it forward. James had made an error in his first equations and had simplified them too much. (Note: There is still some disagreement over the fine points of how the radiometer works. If you want to know more, you can look up "radiometer" on the Internet.) (*4.5)

While he was puzzling over the radiometer, James began having terrible heartburn. He found that baking soda helped to relieve the pain, but would not entirely cure whatever was going on in his stomach. He may have had what we now call a stomach ulcer, the root cause of which is often a bacterial infection. Left untreated, an infected ulcer can turn into a cancerous tumor, and this is what would happen to James. (Antibiotics would not be discovered for another 60 years, and the use of anti- biotics to treat ulcers would not begin until another 60 years after that.) Kath- erine had also begun to have chronic

One of the last portraits

health issues, and James was always more concerned about Katherine than he was about himself.

51

James spent the summer of 1879 at Glenlair, as usual, but did not have the strength to keep up with his normal routines. A friend would later say that "the spring had gone out of his step." In September of that year, one of James's colleagues from the lab came to see him at Glenlair. He knew that James was sick, but marveled at how well he still managed all his responsibilities. James gave his friend a tour of the estate, taking him to all the places he had enjoyed as a child, like the places in the creek where he had paddled around in the washtub. This was the last long walk that James would ever take.

By October James was in pain most of the time and the local doctor recommended that he go to Cambridge to receive treatment from a doctor who specialized in pain management. The Cambridge doctor was able to reduce his pain, but he could not stop the progression of the cancer. The Cambridge doctor said this about James:

"The calmness of his mind was never once disturbed. His sufferings were of a kind that would try the patience of anyone, but they (the sufferings) were never spoken of by him in a complaining tone. In the midst of them his thoughts were rather for others than for himself. His only anxiety seemed to be about his wife whose health had for a few years been delicate and had recently become worse. While his bodily strength was ebbing away, his mind never wandered nor wavered. No man ever met death more consciously or more calmly."

As the Cambridge doctor said, James's biggest concern, even while in great pain, was for the welfare of his wife. He understood that he was going to die and he wanted to make sure that his wife would be well cared for after he was gone. (We don't know what type of chronic illness she had.) James had played the role of nurse for Katherine, receiving instructions from the local doctor.

The local doctor, near Glenlair, said this about James:

"I must say he is one of the best men I have ever met, and a greater merit than his scientific achievement is his being, so far as human judgment can discern, a perfect Christian gentleman."

James rarely wrote about himself, (remember, he disliked autobiographies) but in those last days, he penned these words:

"What is done by what I call "myself" is, I feel, done by something greater than myself in me. The only desire which I can have is like David [king David in the Bible] to serve my own generation by the will of God, and then fall asleep."

The minister of James's church in Cambridge came to see him in the last week of his life. James asked the minister to help him pass the time by reading passages from classic literature and also from the Bible, all of which James had memorized in this youth, or at least had read numerous times, but he enjoyed listening to them again. The minister wrote this about the time he spent with James:

"I had known little about his inner self before his illness. [He was not someone who was inclined to talk about his own spiritual life], although he always attended church every week, and often gave generously to charitable causes. His illness drew out the whole heart and soul and spirit of the man: his firm and undoubting faith in the Incarnation and all its results; in the full sufficiency of the Atonement; in the work of the Holy Spirit. He had gauged and fathomed all the schemes and systems of philosophy, and had found them utterly empty and unsatisfying—"unworkable" was his own word about them—and he turned with simple faith to the Gospel of the Saviour."

Many today would rather not hear about this aspect of James's personality—his deep spiritual life. They like to read about all his equations and brilliant inventions, but feel uncomfortable reading this type of quote. However, to skip over this part of his life would present an unbalanced and unfair view of his life. He wouldn't want us to skip over it.

James died, with Katherine and several friends at his side, on November 5, 1879, at the age of 48, the same age that his mother died, and of the same illness. Katherine lived another 7 years and then she also passed away. But let's end with a happy thought, not a sad one. Let us imagine their great joy when they saw each other again in heaven.

"God will wipe every tear from their eyes. There will be no more death or mourning or crying or pain." Revelation 21:4

It took a long time for the city of Edinburgh to fully recognize the achievement of James Clerk Maxwell. It wasn't that no one knew about him, but there had never been a public monument to officially recognize that this city's greatest scientist was also one of the greatest scientists of all time. His tombstone isn't anything special at all. In fact, he shares the same stone with his wife and parents. There isn't anything to suggest that one of the people buried under it is the man who changed the world by figuring out the relationship between light and electromagnetism, and giving us the equations necessary for inventing all the electronic devices we enjoy today.

Finally, in 2008, a large statue was unveiled at the east end of George Street in downtown Edinburgh. The statue shows James holding his color wheel, and Toby at his feet. One of the side panels shows Isaac Newton doing the experiments that paved the way for Maxwell's work.

By Kim Traynor - Own work, CC BY-SA 3.0, https://commons.wikimedia.org/w/index.php?curid=16281489

James Clerk Maxwell's birthplace, on India Street, was purchased by the JCM Foundation in 1993 and turned into a small museum. Science tools that he made or owned are either here or in the Cavendish Lab. The museum also has many original photographs, letters and manuscripts.

APPENDIX 1: Timeline

1831: Born in Edinburgh on June 13

1839: His mother dies

1841: Started at Edinburgh Academy (a secondary school)

1846: Wrote his first scientific paper: "On the description of oval curves and those having a plurality of foci" which was read to the Royal Society by James Forbes

1847-50: Was a student at the University of Edinburgh

1849: Wrote two scientific papers: "On the Equilibrium of Elastic Solids" and "Rolling Curves"

1855: Wrote his paper on "Experiments on Colour" (the results of his color wheel experiments with red, blue and green)

1856: His father dies

1856-60: Was professor of Natural Philosophy at Marischal College

1857: Won the Adams Prize for his paper about the rings of Saturn

1858: Married Katherine Dewar

1860: Wrote paper: "Illustrations of the Dynamical Theory of Gasses"

1860-65: Was professor of Natural Philosophy at Kings College, London

1861: Demonstrated the first color photograph, and was elected as a member of the Royal Society

1861-2: Wrote paper "On Physical Lines of Force" in which he suggested that light is a form of electromagnetic energy

1864: Presented his famous paper "Dynamical theory of the electromagnetic field" in which he revealed his now-famous equations

1870: Wrote his paper "On Hills and Dales" and published his textbook called "The Theory of Heat"

1871: Made first director of the Cavendish Laboratory, Cambridge, and was its first Professor of Experimental Physics

1870-78: Received several honorary degrees from universities he did not attend (given just to honor his work), was made an honorary member of quite a few scientific clubs in UK, USA, and Europe, and received several notable prizes for scientific research

1877: Published book "Matter and Motion"

1879: Died on November 5

2008: Statue unveiled in Edinburgh

APPENDIX 2: More about cousin Jemima

Jemima Wedderburn (1823-1909) was James's cousin on his mother's side of the family. Her mother was James's Aunt Isabella. Jemima was 8 years older than James but close enough in age to enjoy engaging in childhood games and activities with him whenever their family would visit Glenlair, when James was 5 to 9 years old.

When James started school in Edinburgh at age 10, he stayed with Jemima's family during the school term, so presumably, he maintained a close relationship with her during those early years at the Academy. Jemima was becoming a proficient artist by that time and she encouraged James to learn how to draw.

Jemima married Hugh Blackburn, a mathematics professor who was a personal friend of William Thomson (later known as Lord Kelvin), and who would become another lifelong friend for James.

Jemima would become one of the leading wildlife painters of her day. She specialized in drawing and painting birds. She spent a lot of time outdoors observing the birds before she drew them, so she was able to write about their behaviour, in addition to her artwork. Her first published book was "Birds Drawn from Nature" in 1868. It was immediately very popular and a hand-colored copy was presented to the Zoological Society of London.

Beatrix Potter, famous for her books about Peter Rabbit and other fanciful woodland animals, was a generation younger than Jemima and grew up admiring Jemima's artwork. She received a copy of "Birds Drawn from Nature" for her tenth birthday. It has been suggested that Jemima was the inspiration for Potter's illustrated "Tale of Jemima Puddle-duck."

An owl and a puffin by Jemima Blackburn

APPENDIX 3: More poetry

If you don't like poetry, you can skip this section. For readers who do like poetry, here are a few more samples for you. If you really like poetry a lot, you can read all of his poems in the biography written by Lewis Campbell, "The Life of James Clerk Maxwell." (You can find it at any online book-selling site using ISBN 9798618455046.) This book also contains many letters written to various people throughout his life.

"TRANSLATION OF VIRGIL'S AENEID, 159-169"

(This was a school exercise when James was 13 years old (1844).
His assignment was to translate a section of this famous piece of
Roman literature into English rhymed verse.)

There lies within a long recess a bay,
An isle with gulfing sides restrains the sea,
The waves, divided ere they reach the shore,
Run through the winding bay, and cease to roar;
On this side and on that vast rocks arise,
And two twin crags ascending threat the skies,
Beneath whose shade the water silent lies;
Above, with waving branches, stands a wood,
A grove with awful shade o'erhangs the flood,
And on the further side a cave is shown,—
Within, fresh springs, and seats of living stone—
The nymphs' abode; no chains or anchors bind
The worn-out ships, secure from waves and wind.

"TRANSLATION OF HORACE, Seventh Epode"
This was school exercise from 1846, when James was 15.
He was translating the original Latin into rhymed English verse.

Whither, whither, reckless Romans,
Are you rushing, sword in hand?
Has not yet the blood of brothers,
Fully stained the sea and land?

Not that raging conflagration
Should o'er fallen Carthage play;
Not that the unconquered Briton
Should descend the sacred way.

"Rome," exclaims the joyful Parthian,
"Ruin for herself prepares;
Wolves with wolves are never savage,
Lion lion never tears."

Is this fury? is it madness?
Speedy answer I demand;
Foolish, blinded, guilty Romans,
Silent, stupefied you stand.

Thus, 'tis fated, blood of brothers
Must atone for brothers' guilt,
Since the blood of injured Remus
Romulus in anger spilt.

The first lines of the original Latin poem go like this: (you can find the rest online)
Quo, quo scelesti ruitis? Aut cur dexteris
aptantur enses conditi?
There are a number of ways this can be translated. Many people have made translations, but each translator is free to make their own word choices and rhymes.

Romulus and Remus were the mythical founders of Rome, said to have been raised by wolves.

"A STUDENT'S EVENING HYMN"
(1853)

This is not the whole poem, but selected stanzas that give you a good idea of the feeling for the whole poem.

Now no more the slanting rays
With the mountain summits daily,
Now no more in crimson blaze
Evening's fleecy cloudless rally,
Soon shall Night front off the valley
Sweep that bright yet earthly haze.
And the stars most musically
Move in endless rounds of praise.

While the world is growing dim,
And the Sun is slow descending
Past the far horizon's rim,
Earth's low sky to heaven extending,
Let my feeble earth-notes, blending
With the songs of cherubim,
Through the same expanse ascending,
Thus renew my evening hymn.

Through the creatures Thou hast made
Show the brightness of Thy glory,
Be eternal Truth displayed
In their substance transitory,
Till green Earth and Ocean hoary,
Massy rock and tender blade
Tell the same unending story--
"We are Truth in Form arrayed."

Teach me so Thy works to read
That my faith,—new strength accruing,—
May from world to world proceed,
Wisdom's fruitful search pursuing;
Till, thy truth my mind imbuing,
I proclaim the Eternal Creed,
Oft the glorious theme renewing
God our Lord is God indeed.

"cherubim" is a type of angel
"hoary" means white or gray
"accruing" means gaining
"imbuing" means inspiring or filling up
A "creed" is a statement of beliefs

"I'VE HEARD THE RUSHING"
(1858)

I've heard the rushing of mountain torrents, gushing
Down through the rocks, in a cataract of spray,
Onward to the ocean;
Swift seemed their motion,
Till, lost in the desert, they dwindled away.

I've learnt the story of all human glory,
I've felt high resolves growing weaker every day,
Till cares, springing round me,
With creeping tendrils bound me,
And all I once hoped for was wearing fast away.

I've seen the river rolling on for ever,
Silent and strong, without tumult or display,
In the desert arid,
Its waters never tarried,
Till far out to sea we still found them on their way.

Now no more weary we faint in deserts dreary,
Toiling alone till the closing of the day;
All now is righted,
Our souls flow on united,
Till the years and their sorrows have all died away.

"WILL YOU COME ALONG WITH ME?"

(The poem he wrote to Katherine as a marriage proposal in 1858)

Will you come along with me,
In the fresh spring-tide,
My comforter to be
Through the world so wide?
Will you come and learn the ways
A student spends his days,
On the bonny, bonny braes
Of our ain burnside?

For the lambs will soon be here,
In the fresh spring-tide;
As lambs come every year
On our ain burnside.
Poor things, they will not stay,
But we will keep the day
When we first saw them play
On our ain burnside.

We will watch the budding trees
In the fresh spring-tide,
While the murmurs of the breeze
Through the branches glide,
Where the mavis builds her nest,
And finds both work and rest,
In the bush she loves the best,
On our ain burnside

And the life we then shall lead
In the fresh spring-tide,
Will make thee mine indeed,
Though the world be wide.
No stranger's blame or praise
Shall turn us from the ways
That brought us happy days
On our ain burnside.

*"Ain" means "own." "bonny" means
"beautiful," "braes" means "hillside,"
and "burnside" refers to Glenlair.*

"A LECTURE TO WOMEN ONO PHYSICAL SCIENCE"
(PLACE: A small alcove with dark curtains.
The class consists of one member.
SUBJECT: Thomson's Mirror Galvanometer
(1874)

The lamp-light falls on blackened walls,
And streams through narrow perforations,
The long beam trails over pasteboard scales,
With slow-decaying oscillations.
Flow, current, flow, set the quick light-spot flying,
Flow, current, answer light-spot, flashing, quivering, dying.

O look! how strange! how thin and clear,
And thinner, clearer, sharper growing
The gliding fire! with central wire,
The fine degrees distinctly showing.
Swing, magnet, wing, advancing and receding,
Swing, magnet! Answer dearest, What's your final reading?

O love! you fail to read the scale
Correct to tenths of a division.
To mirror heaven those eyes were given,
And not for methods of precision.
Break contact, break, set the free light-spot flying;
Break contact, rest thee, magnet, swinging, creeping, dying.

This poem shows that Maxwell sometimes wrote amusing poetry about science
or about science experiments. The poem is about a device that William Thomson made
for use at the end of the transatlantic telegraph cable. It could detect very weak electrical
currents. The indicator was a mirror with a bar magnet cemented on the back, which was
suspended on silk threads. A beam of light would shine onto the mirror and was reflected
onto a card with a scale printed on it. The card was placed some distance away from the
mirror in order to magnify the slight variations. This device was used to receive the very
weak signals that came across from America via the undersea cable. Eventually, Thomson
invented the siphon recorder which used a glass tube dipped in ink to record the movements of the instrument so that the message could be printed and then analyzed after it
was received.

"SONG OF THE ATLANTIC TELEGRAPH COMPANY"

This poem was composed by Maxwell as he rode the railway to Glasgow in the year 1857. He had just learned that the transatlantic telegraph cable had snapped while they were laying it across the Atlantic Ocean. According to James, his friend William Thomson had given advice to the engineers laying the cable and they had apparently not followed that advice. This poem was written to the tune of a common song everyone knew at that time, "Over the Sea: Jacobite Song," composed by a Mrs. Groom. (The web address where you can find the sheet music for this song is listed in the bibliography.) He knew that the phrase "Under the sea" would be repeated many times, so he used a math shorthand assigning "u" to be "under the sea," so "2(u)" would be that phrase sung twice: "Under the sea, under the sea"

2(u), Mark how the telegraph motions to me,
2(u), Signals are coming along
With a wig, wag, wag;
The telegraph needle is vibrating free,
And every vibration is telling to me
How they drag, drag, drag,
The telegraph cable along.

2(u), No little signals are coming to me,
2(u), Something has surely gone wrong,
And it's broke, broke, broke,
What is the cause of it does not transpire,
But something has broken the telegraph wire
With a stroke, stroke, stroke,
Or else they've been pulling too strong.

2(u), Fishes are whispering: What can it be?
2(u), So many hundred miles long?
For it's strange, strange, strange,
How they would spin out such durable stuff,
Lying all wiry, elastic, and tough,
Without change, change, change,
In the salt water so strong.

2(u), There let us leave it for fishes to see;
2(u), They'll see lots of cables ere long,
For we'll twine, twine, twine,
And spin a new cable, and try it again,
And settle our bargains of cotton and grain,
With a line, line, line,
A line that will never go wrong.

APPENDIX 4: Suggested Activities

All of these activities are optional, since this is not a curriculum book. However, your student(s) might enjoy doing these activities that correlate with Maxwell's experiments and discoveries. They can be done at the places indicated in the text with an asterisk and number ($*^{1.1}$) or after reading each chapter.

CHAPTER ONE

1.1) This activity is suggested for younger readers, if they are unfamiliar with reflective surfaces. James used a pie pan to "bring the sun indoors." You could use anything that reflects light well, and you can use any strong light source if the sun is not shining through your windows. The challenge is to position your reflective surface so that the light source is shining on it, then adjust its position just slightly so that the reflected light shows up on the ceiling or on a wall.

1.2) Make your own zoetrope! There are many websites, and YouTube videos that show how to make a simple zoetrope. Use keywords "make your own zoetrope."

1.3) Make paper models of polyhedra. You can do your own search for patterns online, or you can try this website: https://www.polyhedra.net/en/
At this recommended website you can click on one of the many models shown, and then below the pictures you'll see links to pdf patterns you can download and print. (This website has many more options than the ones shown on page 7.)

1.4) Draw an ellipse using pins and a string. There are lots of videos showing how to do this on YouTube. Just use keywords to search. Once you get the hang of it, look at the other ellipses on page 8 and see if you can figure out how to draw them. (To do this activity you will need some kind of board, even a piece of thick cardboard, or two pieces of cardboard stacked together, into which you can insert pushpins or thumbtacks or small nails. You might also need to experiment with pencils and pens to find one that works well.)
        Challenge your student(s) to see if they can figure out how to use three or four pins. (These will not be perfect ellipses, but rather ellipse-like shapes.)

1.5) Make some anagrams from your name. Write out your first, middle and last name making fairly large letters on a piece of paper. Cut out each letter separately. Now you can rearrange the letters easily. Can you make other names (or just words) from those letters?

BONUS IDEA: James loved to play with "diablo sticks." which were a new invention is his day. To see them in action, search YouTube or other video service. Some people also post instructions to make them.

CHAPTER TWO

2.1) If you have never experimented with a prism, this is an ideal time to do so. NOTE: Prisms work best in direct sunlight. Other light sources won't give you good results. Also, glass prisms might work better than plastic ones.
        If you can't get a prism, you can try using a flashlight and a glass of water. (A more narrow top will give better results than a wide one.) Fill the glass slowly until it is full, then add a little more, drop by drop, until the water is bulging at the top (view from the side). Then, shine a flashlight sideways through the bulging water. A small rainbow spectrum should appear on the table. If your table is a dark color, you can put a piece of white paper under the glass.

2.2) Experiment with polarized light.

First, what is polarized light? Polarized light consists of waves of light that have been "sorted" by a polarizer, usually a polarizing film nowadays. The polarizing film is in the middle of this diagram, filled with tiny lines. You can see that the light going into it consists of waves pointing in many directions, while the light coming out the polarizer has only one direction. Oddly enough, sunlight gets at least somewhat polarized when it bounces off a reflective surface such as water, an icy road, or the shiny paint on cars. Polarized sunglasses strain out the extra, unwanted light rays, cutting the "glare" coming off water, wet roads, and cars. James only knew about reflected polarized light and minerals like Icelandic spar; he did not have any fancy polarizing lenses like we do today. It is amazing that he could make the discoveries he did without any modern filters or equipment.

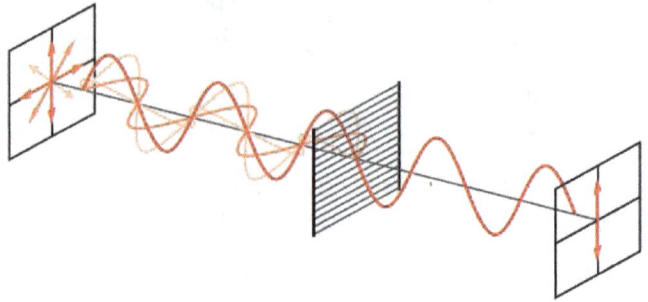

To see colorful stress patterns in clear plastic or glass objects, you will need two polarizing filters. The clear object is put between them, with a light source from behind. There are several ways this can be accomplished.

METHOD 1: If you don't have access to any polarizing filters, you can watch a video demonstration: **https://youtu.be/sKUL2PFLtwY** (Polarized light with sunglasses demo/Ellen McHenrys Basement Workshop-- TheBasementWorkshop channel)

METHOD 2: You can use a pair of polarized sunglasses and an LCD screen (such as a computer screen, or a tablet or phone).

NOTE: Unfortunately, cheap sunglasses might say "polarized" on the label, but they might not good enough for this experiment. You'll just have to try them and see if they work. (If you don't have a pair, or can't borrow a pair, see METHOD 3 for a free way to "borrow" a few pairs for a few minutes.)

You can check to see if your pair of glasses is truly polarized by holding them in front of a blank, white LCD screen (open a blank word document on a table or computer) and then rotating the glasses. LCD screens have a polarizing filter built in, so you can use the screen as one of your filters. The view inside the glasses should alternate between light and dark as you rotate the lens. When the two polarizing filters are lined up in the same direction, light will be able to go through. When the lines in the filters are perpendicular (at 90 degree angle) no light will get through and the glasses will go dark. You won't be able to see the tiny polarizing lines in your sunglass lenses. They are far too small—truly microscopic.

Now you will need some clear plastic or glass objects. Clear plastic silverware works well, as does clear tape and the clear plastic housing around the tape. You can try clear plastic food wrap—this won't be as colorful, but you can see dark color spots appear as you stretch it. But use whatever you have available.

Put your clear objects between the sunglasses and the screen. You should be able to see some brightly colored patterns (as you rotate the glasses), similar to the image shown here. The shapes in the pattern will indicate where the stresses are in the material. It might also show up cracks that you can't normally see. If the plastic or glass was stretched or squished while it was being made, the patterns will show this.

It will take two sets of hands to do this, but try holding a piece of clear tape or plastic wrap between the filters and watch it as you stretch it. The patterns should shift and change as you watch. Another interesting thing to do with tape is to cut many pieces and layer them at angles across each other, then put this between the polarizers and turn one of them. You will see the color patterns shift and change in beautiful ways, almost like a stained glass window.

If you want to record what you are seeing, you can put the sunglass lens over the camera lens on your smart phone or digital camera.

METHOD 3: "Borrow" some sunglasses at a store

Find a store that sells quality sunglasses and be a shopper who wants to try them out. Be sure to try on a few pairs so you look like a real customer. (Big stores that sell many other things probably won't even notice you.)

NOTE: You will need to be very careful when doing this. You don't want to touch the lenses with your fingers—don't get your greasy fingerprints on their merchandise! And don't drop them.

You should have a few small clear objects in your pockets that you can pull out to put between the lenses of two pairs of glasses. (This will require a second pair of hands.) Rotate just one of the pairs and you should see the same thing you would see using METHOD 2, except that your field of view will be smaller.

2.3) Try drawing a cycloid shape

You will need a piece of paper, a piece of cardboard, a pencil, a ruler, an object with a round base (such as a drinking glass), a pair of scissors, and something with which you can poke a hole in the cardboard (a sharp pen point or possibly the scissor tip). Set the round-base object on the cardboard and use the pencil to trace around the base, leaving a circle on the cardboard. Cut out this circle. Now poke a small hole in the very edge of the circle. Put the ruler (or any straight edge) onto the paper. If it slips around you can secure it with tape. Put the cardboard circle on top of the ruler. Put the pencil point into the tiny hole so that it will trace a line on the paper. Carefully roll the circle across the ruler. This will be tricky! When finished, you should have a cycloid shape drawn on the paper.

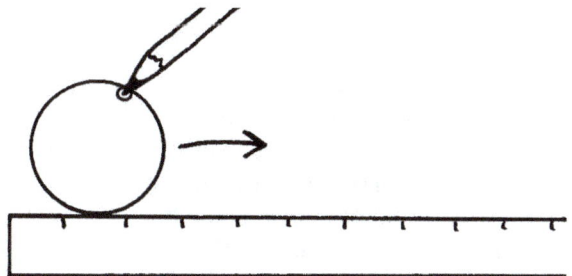

2.4) Watch people dropping cats. Yes, really! You can even find slow-motion videos showing exactly how cats turn as they fall. Go to a video service such as YouTube and use keywords "Why cats always land on their feet."

2.5) Find the blind spot in your eye. You can find lots of online instructions, but try this one from the Exploratorium: https://www.exploratorium.edu/snacks/blind-spot

2.6) You can prove to yourself that James was right when he said that the colors we see in nature are combinations of red, green and blue light.
(This is different from paint colors. With paint, colors are a combination of red, blue and yellow.)

PART 1: To see red, green and blue light combine to make colors, you will need a magnifying glass and any type of screen: a computer, tablet, or a phone. Pull up a color photo on the screen and then look at the colors under magnification. Look at many colors, especially yellow, white, green, orange and brown or black (anything but red, green or blue). What do you see? Red, green and blue pixels under every color, even white! It's amazing!

PART 2: Here is another color activity you might want to try. Use a compass to draw a circle on a piece of cardboard, and then several more circles of the same size on a piece of white paper. The exact size is not important but make sure the circle is at least 10 cm in diameter. Cut out all the circles. Push a sharpened pencil through the center dot in the cardboard circle (where the compass pin was) with just a few centimeters of the pencil sticking through the bottom side. This will create a spinning top. Then place various colors on the paper circles. You can use color paper shapes, or a marker or crayon if you make the shape solidly colored. If you want to do something most similar to what Maxwell did, cut whole circles out of colored paper, then cut a slit in each one going from the outside to the inside (just one slit in each circle). This will allow you to overlap several circles (by slipping them into each other's slits) and adjust how much of each circle is showing. You will have to secure them to the cardboard circle each time, but they will stay in place with just a tiny bit of tape, an amount easy to remove each time.

Make a hole in the center of the paper discs and place them one at a time on top of the cardboard disc. (If they slip you can anchor them with tape.) Spin the top and observe what your eye perceives as the colors spin. If the pencil seems to be slipping and not sticking to the disc, secure it on the bottom with a bit of tape or some hot glue. The pencil should be sticking straight up, perpendicular to the disc. If it is tilted, the disc won't spin well.

TIP: If you want the top to spin for a longer time, you can tape coins to the bottom of the cardboard circle. The extra weight will give it more spin time.
NOTE: Keep this top for the first activity in the next chapter.

2.7) Observe magnetic lines of force.

You will need: a magnet (a rectangular ceramic magnet, NOT a horseshoe-shaped magnet), a piece of paper or white card stock, tape (masking tape is fine), a pencil, a pin, and a piece of thread   Optional: steel wool (see notes below)

Faraday used iron filings (pieces of iron as small as grains of sand) to show the lines of force, as shown on page 23. You could use homemade iron filings by squeezing and shaking a piece of steel wool over a piece of paper, but this does not come without risk. The tiny bits of steel easily get stuck in the skin causing irritation, and they are a great danger if they get into your eyes. (Tiny pieces of steel in your eye can make you unable to get an MRI for the rest of your life.)

Instead of tiny bits of iron, you can work with something larger: pins. Pins are sharp and have to be used with care, but they are a lot safer than iron filings.

First, tape the magnet in the middle of your piece of paper. (See illustration below.) Then, tie the thread to the middle of the pin, then make sure that the pin balances parallel to the table. Now touch the head of the pin to one end, or one side, of the magnet. This will magnetize the head of the pin.

Now dangle the pin near the magnet in various places but don't let it actually touch the magnet. When you hold the pin close to the magnet, hold the thread down close to the pin so the pin can't swing around. For positions further away, it is less important where you hold the pin. Make sure the pin is free to swing into any position, because the pin will show you the direction the magnetic field lines. Take note of which direction the head of the pin is pointing.

To record what you are seeing, use the pencil to draw a line right under the pin. Make sure to indicate which direction the head is pointing. You can draw a tiny short line to represent the head. You will need to do this dozens of times before the all the lines of force become apparent. A sample of what it might look like is shown here. Yours might not look exactly like this, but similar.

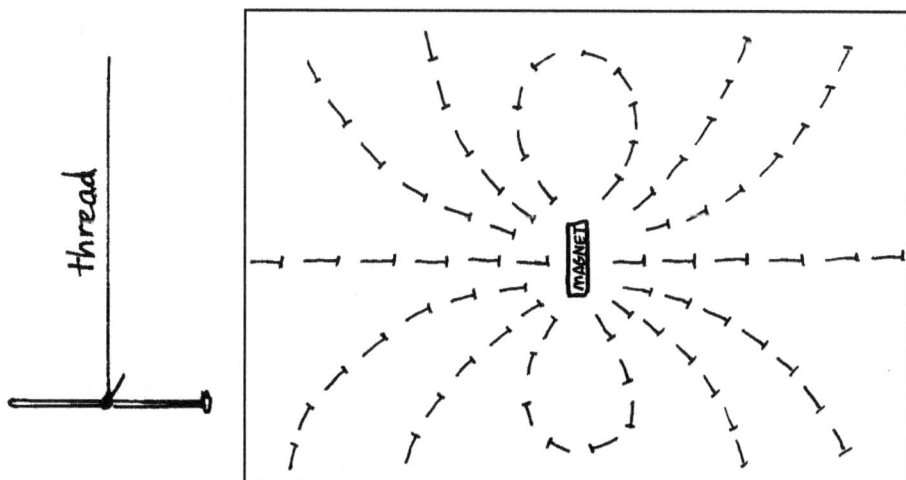

CHAPTER THREE

3.1) Make a simple model of Saturn's rings.
You will need the cardboard top that you made in activity 2.5.

Cut a paper circle the same size as the cardboard circle. Use a marker to put dot onto the paper. These dots will represent the rock and dust particles that surround Saturn. Make more dots in some places, and fewer dot in other plac-es. Now put the paper onto the cardboard circle. (Punch a hole in the center of the paper and slide it down the pencil onto the top of the cardboard disc.) Spin the disk. Do you still see individual dots while the disc is spinning? Try making another pa-per circle and use a different arrangement of dots. You might want to make some dots just on one side of the disc. Spin this and observe what you see. Does this explain why we see Saturn's rings as con-tinuous discs? Does it explain why there was doubt as to what these rings consisted of, whether solid, liquid or particles?

3.2) Make a simple model of gas diffusion. You will need a cereal box (or a similarly shaped flat box), scissors, tape, and some type of small spheres: mar-bles, very round dried beans, BB's, plastic ammunition pellets, or even spherically shaped pieces of cereal (like Kix® in the U.S.)

Cut off one of the larger sides. Use this cut-off piece to cut a divider that will go across the middle of the box. Cut a hole in the divider just a little bit larger than the size of whatever type of sphere you are using, then tape the divider into place. Put the spheres into one side of the box. The spheres will represent gas molecules. The hole is where they can escape into the air. Shake the box, creating sudden, random motions so that the spheres are rolling around very fast in all directions. Air mol-ecules move very fast in the air. As you watch, a sphere will get through the hole from time to time. Once there are an equal number of spheres on each side, it will basically stay this way. Real gas molecules behave in a similar way.

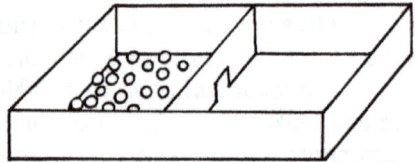

3.3) Static electricity. It takes a lot of static electricity to make your hair stand on end like in the photo on page 31, but you can do a small experiment with it. Your student(s) may have already done simple experiments with static, such as charging up a balloon by rubbing it on their hair and then sticking it to a wall or the ceiling. (This works best when the weather is dry, or if the humidity is low in the room you are in.) However, have they ever made static sparks in the dark? It's much more ex-citing to actually see these tiny bluish-purple sparks! (The room needs to be really dark for this.) Sometimes the sparks are surprisingly long. A wool sweater rubbing against a plastic laundry basket can really get some sparks flying! If you need some extra static experiment ideas, the Internet is full of them. Classic demos include "separate salt from pepper using static," "bend a stream of water," and "move an empty soda can."

3.4) Watch a short video about the history of color photographs.

It would be very difficult to construct an activity that would simulate what Maxwell did in his famous color photograph demonstration, so, instead of a hands-on activity, you might just want to watch a short video that gives a little more background on what was happened before and after this. Search YouTube for "How Color Photography Was Proven to be Possible" by the Filmmaker IQ channel. The video is about 8 minutes long.

3.5) An activity about the first observation about electricity and magnetism (in red on page 36)

This activity focuses on the first of the four observations:

1) "Like" forces repel and "unlike" forces attract. These forces follow the "inverse square law" meaning that as the distance increases arithmetically (1 2, 3, 4, 5...) the force decreases geometrically (1, 1/4, 1/9, 1/16. 1/25...)

The first part is not hard to understand. Kids catch on very quickly that magnets have two poles, north and south, and that opposite poles attract and "like" poles repel. The second part seems hard to understand because it involves numbers. Here is a way to experience the meaning of these numbers without having to do any difficult math.

You will need a magnet, 15 paperclips, a pencil, a pen, a piece of graph paper (or improvise with regular paper), and a long piece of masking tape.

First, cut 60 pieces of masking tape that are about the size of a postage stamp. Label them with numbers 1 to 60. This will help you keep track of how many pieces are in your stack during the experiment. You won't have to worry about forgetting how many you put on.

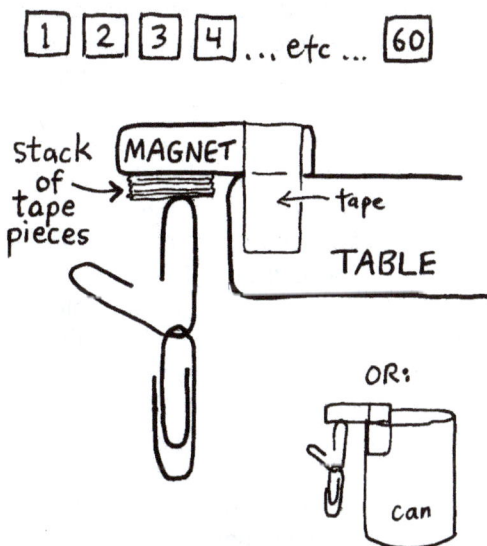

Next, tape your magnet to the edge of a table so that one end is hanging off the table. Now open one paper clip so that it forms a hook on which you can hang other clips.

Before you start the experiment, get your graph ready to record your results. (If you don't have graph paper, you can improvise using a ruler to draw the horizontal and vertical axis lines, and then some very light lines going up at the 10, 20, 30, 40. 50 and 60 lines on the horizontal axis, and over at the 5, 10 and 15 lines on the vertical axis.) See the sample on the next page. (You are welcome to photocopy it.)

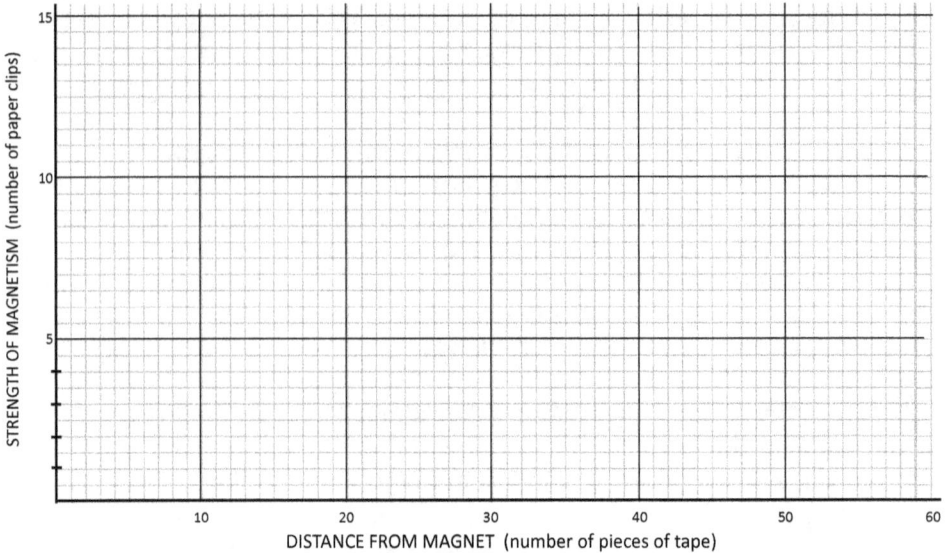

Start with no pieces of tape on the underside of the magnet. Hang the hook clip, then count as you add paperclips to the hook. When the hook falls, the last number of clips it held successfully is the number you will record on your graph. The first dot will correspond to the zero on the bottom line and probably about 12 (maybe one more or one less) on the vertical STRENGTH line. Then take the hook off and add the piece of tape labeled 1. Then try hanging clips again. Record the successful number of clips it held. Then add piece of tape number 2. Repeat the hanging test. Then add piece of tape number 3, then 4, then 5. After 5, add numbers 6, 7, 8, 9, and 10 all at once. So your next trial will be with 10 pieces of tape. Then add another 5 pieces, bringing you to tape number 15. Then add 5 more, bringing you to tape number 20. Then start adding 10 pieces at a time: 30, 40, 50, and 60. Record your successful number of hanging clips for each trial.

The expected result is that your graph will look similar the one below. Notice how steep the line is between 1 and 10 pieces of tape, and how it gradually gets less steep, almost flattening out, by 60. The is the visual representation of "The Inverse Square Law." The attractive force of the magnet drops off quickly at first, then more slowly. If you had professional equipment and could take precise measurements, you would find that the force decreased inversely with the square of the distance between the magnet and the object. If the distance is twice as far, the force will be $1/2^2$, or 1/4 as strong. If the distance is 4 times as far, the force is $1/4^2$ or 1/16 as strong. Our experiment wasn't precise enough to derive this law, but it at least gives the general idea of how the force decreases over distance.

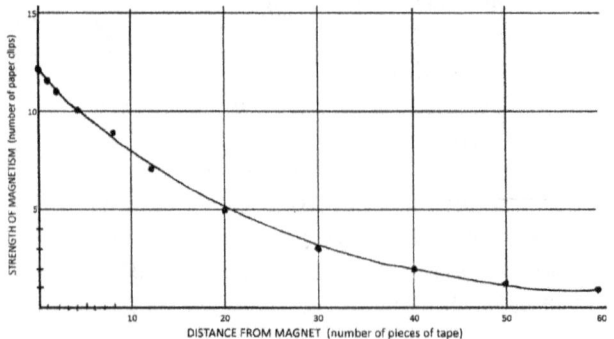

CHAPTER FOUR

4.1) Make a simple torsional pendulum. You will need the cardboard disc that you made in activity 2.5. (Or make a new one: use a compass to draw a circle on a piece of thick cardboard. Make sure the circle is at least 10 cm in diameter, and make sure the central point made by the compass point is visible.) You will also need a piece of tape, at least 15 cm long.

Fold your piece of tape in half the long way. This will take some patience. If the fold gets messed up, just cut a new one and try again.

Push one end of your folded tape through the central hole in the cardboard disc. If the hole is not large enough you can use a pencil point to make it a little larger. The end of the folded tape should stick through about 1 cm. Use a small piece of tape to secure that bit of tape to the bottom of the disc.

Now you will be able to hold the top end of your folded tape with the disc dangling at the bottom. It should be hanging approximately parallel to the floor (flat, not at an angle). While holding the top of the folded tape, give the disc a gentle spin. It should turn several complete revolutions, then stop and switch directions, spinning the other way. Then it will stop and again and then start spinning the other direction again. Back and forth, back and forth, the disc will keep switching spin directions until it runs out of energy and stops.

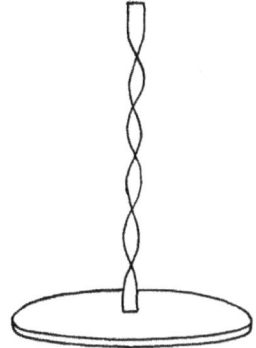

Maxwell's disc was much heavier than your cardboard disc. This did not make it spin faster, but it did help it to spin for a longer time. Your disc will stop spinning in about 30 seconds or so. (If you disc is larger it might spin longer.) The energy that drives the rotation of the disc is the twist in the folded tape. Maxwell used a metal strip, not a plastic one, but his pendulum worked basically the same way that yours does.

4.2) A video explanation of resistors and Ohm's Law

A hands-on experiment about resistors would require more than just objects you have lying around the house. We would need wires and batteries and volt meters. So here is a great 4-minute video giving you a really good, understandable summary of how resistors work and why they are used in electronics. Both the graphics and the narration in this video are exceptionally good. "Resistors Explained & Ohm's Law Made Simple! | What Are Resistors? & How Do Resistors Work?" on the YouTube channel called "Engineeringness"

Here is another fabulous video to watch: "MAKE presents: The resistor" (This is by the "Make:" channel. They also produce the famous "MAKE" magazine.) At the end of this video they show you a simple experiment to try if you happen to own a digital multimeter. If you don't own one, you can enjoy watching the experiment.

If these videos are no longer available, try searching using key words "Why resistors are used" or "How resistors are made."

4.3) Turn electricity into magnetism. Your student(s) may have already done this in an earlier grade level. However, more complex experiments about electricity and magnetism are beyond the scope of this book.

You will need: a battery, a long piece of copper wire (it can have insulation around it, no need to strip the wire except at the very ends), a large nail, and some paperclips.

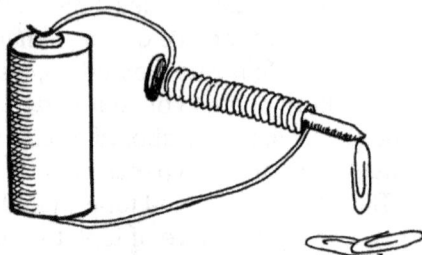

Wrap the wire tightly around the nail, leaving the tip of the nail showing. Make sure there are "leads" of wire at both ends, which will attach to the battery. The ends of the wires do need to be bare, so if the wire is covered in plastic insulation you will have to strip off a bit.

First, check to see if the nail has any magnetic properties before you attach the battery. Touch the tip of the nail to some paperclips. Unless the nail was previously magnetized, the nail will not have any magnetic properties. Now touch the ends of the wire to the terminals on the battery and try it again. Does the nail attract the paperclips? The nail should be able to hold quite a few clips. Now remove the wire from just one end of the battery. Do the paperclips fall off? The nail will only be magnetic while the wires are in place on the battery.

This set up is called an electromagnet. As Maxwell stated, electricity and magnetism go together. When you have one, you have the other. When there is electricity (from the battery) flowing through the wire, this will create an electric field. When the flow of electricity stops, the magnetism vanishes.

There are ways to show the reverse, to use magnetism to create electricity, however, the set up is not as simple as this one. If you are interested in knowing more, you can search the Internet for "experiments using magnetism to create electricity."

4.4) "Maxwell's Demon" (a thought experiment about entropy)

While James was busy thinking about the behavior of gas molecules, he thought of a funny demonstration that he could do only in his mind, not in a lab. He invented a little imaginary character that would later become known as "Maxwell's demon." (The use of the word "demon" here does not imply anything evil. Here, the word means something like a fairy or an elf. It was not James himself that called it a demon; he called it a "finite being." It was William Thompson who called it a demon.) This imaginary being was only slightly larger than the gas molecules themselves. The imaginary jar in which he lived was divided in half. Between the two halves was a wall with a tiny hole just large enough to let one molecule pass through. (Think of the box you constructed in activity 3.2.) Over the hole was a little shutter that the "demon" could open and close. Every time a very slow moving molecule in section B would come close to the shutter, he could open it to let the molecule pass into side A. When a very fast-moving molecule in section A came by he could open the shutter and let it pass through into side B.

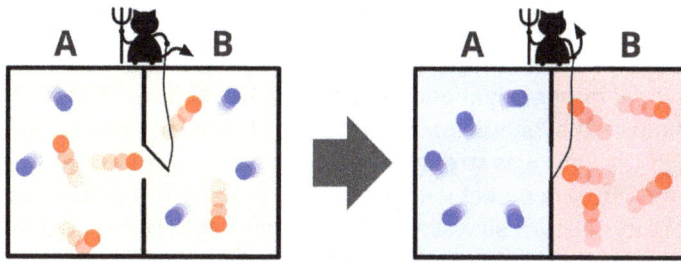

The demon controlled the back-and-forth movements of the molecules so that all the slow molecules were eventually collected into one side and all the fast molecules into the other. Thus, the demon as able to make natural laws act in reverse. In nature, when you start with something hot and something cold, the hot thing cools off and the cold thing warms up, so that they end up being the same temperature. If you pour hot water and cold water together, you'll end up with lukewarm water that is neither hot nor cold.

When something goes from being highly ordered (hot and cold separated) to being less ordered (all mixed together and lukewarm) we say that "entropy" has increased. The concept of entropy applies to many things, even to a room that starts out neat and tidy then gets messy as the days go by. It takes energy to put the room back into order. Physicists say that the whole universe is getting less ordered as time goes on. Energy starts out in a useful form (fuel of some kind) then gets used up as it is turned into motion or heat which may be useful right at the moment, but can never be recovered and recycled. Once the car has traveled down the road, you can't get the gasoline (petrol) back again. However, in this thought experiment, Maxwell has found an imaginary way to decrease entropy, not increase it. The jar starts out at equilibrium (both sides the same temperature) and then ends up with a hot side and a cold side.

James then explained to his readers that information about gases is statistical, and describes only the behavior of the gas as a whole, not the individual molecules. Even a tiny demon the size of a gas molecule would be unable to know enough about the movement of each molecule to be able to sort them. Even though the thought experiment "failed," it still provided an interesting teaching tool.

4.5) Black versus white   It would be too hard to make a radiometer since it requires the manufacturing of a bulb with no air inside it (a "vacuum"). However, you can still do a simple experiment that demonstrates how dark and light colors are affected by light.

You will need two items, one dark and one light. These can be a black piece of a paper and a white piece of paper, a black (or very dark color) shirt and a white shirt, or any other items made of the same material with only a difference in color.

Set the objects in sunlight, either outdoors or indoors if the sun is coming through a window. (If sunlight is not possible you might be able to achieve the same effect with a very bright bulb, like a spotlight bulb. LED bulbs probably won't give off enough light and heat.) Leave the objects in the light for at least 15 minutes, maybe longer. Then lay your hands on them to feel their temperature. If the light was strong enough, you will feel quite a difference between them. The dark object will be noticeably warmer than the white one.

Dark colors absorb all wavelengths of light—that is why they are dark. White objects reflect all wavelengths of light—that is why they are white. Light waves are made of energy, so the dark objects absorb more energy. The energy affects the molecules in the objects, causing them to move around more. Molecular motion IS heat, so the dark object has more molecular motion and thus more heat.

* *

AN ADDITIONAL VIDEO ACTIVITY: The Dynamical Top

During the time that James was at Marischal College and still experimenting with his color wheel, he also developed a very sophisticated spinning top that he called the "dynamical top." We wrote about it in a letter to his father, so we know that he invented it during or before the year 1856.

This top was designed to show "the motion of a system of invariable form about a fixed point." It was similar to a gyroscope that stays upright while spinning. He could stick his color wheel to the top of it and study the subtle motions of the top by looking at what happened to the colors.

To read a description of the top, you can download a pdf written by Roger Clark of Aberdeen University. In this paper you can see Maxwell's equations describing the motion of the top. He had a gift for being able to express physical observations with math.

He had a brass model of the top made in 1857 and exhibited it at the Royal Society of Edinburgh in that same year. **You can see a video of a facsimile (a copy) of this top by going to YouTube and search for "Maxwell Top" by the "gyroscopes" channel.**

BIBLIOGRAPHY

BOOKS:

Campbell, Lewis, and William Garnett. The Life of James Clerk Maxwell. Original copyright MacMillan and Co., London, 1882. Copyright renewed by Sonnet Software, Inc., 1997 and 2020. ISBN 9798618455046.

Mahon, Basil. The Man Who Changed Everything: The Life of James Clerk Maxwell. Wiley, 2004. ISBN 978-0470861714

WEBSITES:
https://homepages.abdn.ac.uk/j.s.reid/pages/Maxwell/Legacy/MaxTop.html
https://clerkmaxwellfoundation.org/
https://atlantic-cable.com/Article/UnderTheSea/index.htm
https://collection.sciencemuseumgroup.org.uk/objects/co32961/
      thomsons-mirror-galvanometer-1858
Sheet music for "Over the Sea":
      https://babel.hathitrust.org/cgi/pt?id=mdp.39015096603868&seq=2

Wikipedia articles:
James Clerk Maxwell
John Clerk Maxwell
Lewis Campbell
P. G. Tait
Michael Faraday
Jemima (Wedderburn) Blackburn
Maxwell's Demon
Polarized light

VIDEOS (from YouTube):
"BBC Scotland: James Clerk Maxwell" (use these keywords to search)
"Maxwell's Amazing Spinning Top" (Grand Illusions channel)
"James Clerk Maxwell- A Sense of Wonder- Documentary" (Short Form Docs)

www.ingramcontent.com/pod-product-compliance
Lightning Source LLC
Chambersburg PA
CBHW071420040426
42445CB00012BA/1228